"BEYOND THE LABELS"

'CREATIVELY GIFTED'..........A MOTHER'S STORY

A GIFT
to Holism

A gifted young woman's journey to adulthood as she passes through agorophopia, chronic fatigue and "hidden disabilities" to wholeness.

ISBN 0-9647794-0-4

Distributed by:
Karma Publishing
PO Box 16103
Rumford, RI 02916

Printed in the United States of America

DEDICATION/ACKNOWLEDGEMENT

I dedicate this book to my two children for plodding through much adversity;

to all those who have had faith in me; for all those who have not understood or had faith in me - you, too, are a part of me;

to all those who may relate to what is written, you have my support and encouragement to be who you are and to be happy and confident in your own personhood.

Sincerely,

Elaine

October 14, 1995

Dear Robert
and Ellen
Best wishes to you
in your many endeavors:
Good fortune and a happy
future with your daughter,
Bridget.

Sincerely
Elaine

"God rests in reason" and "moves in passion". (Gibran)
(With special thoughts of Jeanette Tamagini)

Preface

This is a book about love, courage and people. It is a book about labels and the connotations that accompany them; Finally, it is a journey to holism. It explores, and ultimately asks us to understand and celebrate, the nature of differences - biological, biochemical and other "differences".

Beyond The Labels walks the reader through the conditions known as "Giftedness", "Agoraphobia", "Chronic Fatigue" and other "hidden disabilities". With a combined knowledge of anatomy, physiology, and the emotional complexities of the human heart, it sheds some light on the implications of what these terms might imply and what others may carelessly call the manifestations of them. It emphasizes the importance and fragility of our interactions with one another and with the environment that surrounds us.

While this story depicts the experience of one family, it reflects the common and constant struggle to achieve integration within ourselves and with our surroundings. It also describes some of the obstacles present in our society which deter or even halt the actualization of our inherent potential for wholeness.

This is a book for people who do not understand themselves, for people who are not understood by others in society and for others who may even have been misled, limited or "imprisoned" spiritually, intellectually and/or emotionally, by the use of labels. I hope that through reading this story and the accompanying lessons, that some of you will be able to recognize and appreciate our very innate human differences and with that knowledge, move you closer to the "Kingdom of God" that is within you and in which we all share.

Contents

Introduction ix
A Nursing Career xv

The Story
From Nursing Career to Family Career 3
The Birth of an Angel 7
The Arrival of a Brother 15
"Real" Problems of a "Real" Family 21
Ready for School 25
Nutrients, Allergies and "Stress-related Conditions" 29
"Branching Out" 33
A New Pediatrician - A New Perspective 37
The New Label "Giftedness" 41
Ties to Tradition 47
A Time for Mom - The Beginning of "Holism" 53
A Program that Works 57
Talent Develops 61
On to Junior High 67
A Camping Experience - A Turning Point 71
Where is my Daughter? 77
What to Do About School 81
Another Look at Physical Aspects 87
Another Facility.....Another "Treatment" 91
Some Real "Clues" 99
What is this Thing called "Giftedness"? 107
Success Comes in Small Packages 113
Disaster Returns 115
A New Therapist, A New Vision 121
The Beginning of a "Miracle" 133
On to Senior High 139
Senior Year -
The Onset of Chronic Fatigue, Another Attack 147

Contents (cont.)

College, a Choice....Whose Choice? 157
Finally, An Integrated Accurate Assessment 161
Evaluating the Devastation 165
Beginning the Journey Back 169
College "Rules" 171
Growth "Happens" 175
A Legacy 179
Mom and Karen "On the Move" 181
Return to Home 185
Life Goes On 187
The Blooms of Spring 193
Home 197

Panic Disorder 205

Transition to Knowledge 217
"Hidden Disabilities" 219
Chronic Fatigue 220
Analysis 221

Lessons
"Hidden Disabilities"/Individual Differences (Their Meaning) 227
"Giftedness" and "Potential" 233
"The Battle of the Sexes" 251
Morality and Conscience (as related to "Gifted") 259

Conclusion 267

Further Readings 275

Order Form 285

Introduction

Beyond The Labels: "Creatively Gifted"...A Mother's Story, is a heart-wrenching and heart-warming account of a family's struggle to raise their two "gifted" children to adulthood. It tells of the mother's impassioned plight to get professionals to listen to and consider that inherent biological and biochemical and other innate differences can be causative factors in some of the behaviors that they often too quickly stereotype, characterize and label. By virtue of the story that it tells, it has much knowledge to contribute to society. In relating the journey, it portrays some of the characteristics associated with those who are often called "gifted" and how those characteristics can work as liabilities or assets. It shows society's general and basic lack of knowledge about the nature of some of our "differences", without which they cannot help many to achieve their true potential.

There is much emphasis today on standards, physical norms, I.Q. test scores and the so-called "pursuit of excellence". But I believe that not enough emphasis is placed on the pursuit of understanding and moving toward the creation and/or fulfillment of our own unique potential. I emphasize unique, because the attainment of our potential has to include our own basic biochemical, genetic and spiritual endowments. We all vary in that and in the extent and type of sensitivities that we experience with other people and to the environment.

This story tells of what can happen to people and families when "helping" institutions are not equipped with sufficient knowledge and/or experience about human differences that is needed if they are to truly help others. In our case, because of many professionals' limited knowledge, and their attempts to control, values and behaviors that we viewed as positive were seen by others as negative characteristics and were to be "treated". In essence, those characteristics were the "stuff" that creativity and potential are made of. In the process of those "treatments", a family became scattered and scarred.

Much has been said recently about the difficulties that bright young women encounter in school and other settings, as they struggle to maintain their identity. But these struggles probably never end. *Beyond The Labels* portrays the continuous spiritual/moral, journey people must face if they are to remain true to themselves in today's society. It reflects the journey of the gifted children's mother, who, herself, had often been condemned for her values, but had succeeded in spite of that. Prior to marriage and child-rearing, she had many worthwhile and notable accomplishments. But while bringing up her children to develop their own unique talents, she was presented with many obstacles, including the threat that her daughter could be taken from her because prolonged, numerous absences were in non-conformity with school regulations.

But this is not a story of rebellion. It is a story of the journey of two parents, each with their own differences, trying together, to raise a "gifted" male and female.

Both children demonstrated talents and high motivational levels early in life. Both met several criteria by which the school system characterized them as "gifted". By third grade, a time when screening was done for gifted programs, school personnel had noted many outstanding characteristics which qualified their young daughter as "gifted":

- Two grades above level in reading,
- An individual I.Q. of at least 130;
- A high level of social and emotional functioning - and other notable features.

She was referred to and attended a program for the "gifted". That program met her needs and nurtured her well in grades four through six. But what happened after that is the crux of the story. Although the story relates the different routes taken by the children as they developed their own uniqueness and talent, it particularly emphasizes the struggle of the young girl to become the "gifted" young woman that she is today.

During adolescence, she had a very difficult time. Although she strived intensely to maintain the self-identification which she knew was hers, she was not being adequately nourished and affirmed through the systems in place to provide that.

Following her years in the gifted program, she was placed in a program that emphasized acceleration and resulting scholastic scores. In high school there was a much stronger emphasis on the "required" norms and grades for college entrance. Scholastic performance, of necessity it seemed, had to conform to certain academic standards and to the expectations of those who taught them. The development of creativity for creativity's sake could be done on an extracurricular basis, but certainly did not seem to be promoted in a rigorous program that seemed only to emphasize acceleration in preparation for college. This was very different from the expansiveness and initiatives allowed in the earlier programs for gifted pupils, particularly for those who had a penchant for creativity. These factors seemed to make the adolescent struggle even more difficult. It became further complicated by the long-term history of environmental and chemical sensitivities as they became aggravated by the ongoing physical changes of puberty.

During this period, the symptoms of her long-term childhood "allergies" developed into different ones which are now called "panic attacks", "agoraphobia" and "chronic fatigue". These symptoms were not understood then, and therefore were not identified accurately nor were her needs met. They resulted in long-term absences from school. Professionals from whom help was sought, applied their own limited knowledge base and formed firm conclusions, not being open to the fact that there might be other reasons for the absences, nor that her mother might have had some significant experiences that gave her observations credibility. In fact, blame was even projected onto the mother as the problems increasingly were portrayed to be caused or aggravated by an "over-protective" mother with "family problems". The parents were even told that a court case could be made to have their daughter taken from them if she did not meet the school attendance requirements. The "treatments" to accomplish this were horrendous.

The complexity of these factors and their interactions on a family are told in story form. It allows the reader to follow the journey as the mother experienced it and tells it. It discusses the factors and values of society that somehow became prominent issues in the course of the journey, important ones which can help or hinder the development of one's potential; factors such as: what constitutes "giftedness"; male/female roles and interactions; views on conscience, morality and the "Six Stages of Moral Development"; and biochemical, biological and "other" differences. It asks thought-provoking questions about the sometimes strange rules of our society and gives insights about the nature of "potential".

Fortunately, this family's story has a successful outcome. For, later on, both mother and daughter encountered professionals who had the maturity, knowledge and understanding that it takes to confirm and foster individual potential.

The young woman is now a Phi Beta Kappa (summa cum laude) graduate of an esteemed private college in the Northeast and has had many successful accomplishments. She has already co-authored a chapter in a leading professional teaching guide, with a professor who has termed her "among the best and brightest". She teaches full-time and clearly demonstrates her "giftedness". Even though she is still affected by various environmental and food sensitivities, she has learned how to deal with the "quirks" of her sensitive immune system, having developed her own proven beneficial methods of holistic practices and philosophies into her lifestyle.

The "Mom", a successful nurse prior to her marriage and child-rearing, now has a private practice as a nurse-counselor. She educates, advocates and helps other individuals and families who suffer from some of the same problems, circumstances and inappropriate "categorizing" which her family experienced. She is involved in many professional organizations, including that of the International Honor Society of Nursing and is a now state coordinator for the American Holistic Nurses Association, emphasizing and honoring its philosophy of integrated wholes.

Beyond The Labels emphasizes attainments by looking at life and problem-solving in a holistic manner. It asks the institutions of Western society to further their understanding of people who do not "fall into" their present categories of knowledge and whose reality may well be different from the commonly accepted "norms". It champions the advocacy and understanding of the many aspects of our unique differences and potential "giftedness."

Please read this story to open your heart to a room full of love....and knowledge.

A Nursing Career
(Professional, Caring and Knowledgeable)

The 1960's were the years of rebellion and individuality. I was a young graduate nurse with a promising career and successful start in the field of Public Health Nursing. That particular aspect of nursing appealed to me because of its emphasis on seeing a patient/client within a perspective of the whole. Patients were viewed and treated as individuals within a family unit, within a community, and, beyond that, to the larger social system. My role was to visit individuals and families for the purposes of carrying out bedside nursing, teaching various aspects of disease prevention, and promoting health.

I had done well from the start and was given increasing responsibilities. I initiated one of the many child health clinics in the state where I worked cooperatively with pediatricians in assuring that young infants had their required immunizations; that mothers understood aspects of formula preparation, safety precautions, and the control of communicable diseases. We, staff nurses, talked with families about their problems, sibling rivalry, ill parents and whatever else that might be involved in promoting the health and well-being of new mothers and their infants. I enjoyed the work very much and was happy when I learned that our agency would participate in a special research study of newborns and their development that was being conducted by a nearby well-known university. Educational meetings were provided for the staff nurses involved, whose role was to make observational and instructive home visits as part of the study. The medical staff from the research project reinforced our knowledge about infant growth and development and informed us about what was thought to contribute to a "healthy" infant and what signs to look for that might indicate potential problems. It was part of our job to report our findings. So when we visited newborns and watched their growth, we did so with a "trained" eye. We carefully scrutinized the infant's physical conditions, alerting the physicians to anything unusual or suspicious that we thought should be brought to their attention. We examined leg folds and condition of eyes; monitored hand-eye coordination and other behaviors. I was very "tuned-in" regarding what conditions warranted watching, and I eagerly applied the knowledge that had been presented by the staff of that Child Development study.

Within a few years, I found myself employed in a Civil Service position in the nursing division of the state Department of Health. My work responsibilities in the field of preventive health care greatly expanded. In addition to the duties and functions that I had carried out while in the Visiting Nurse Agency, many more were added, particularly in the field of maternal-child health, even including some part-time school nursing.

I was assigned to rural areas of the state where I was able to experience the fascination of working closely with Native Americans. I loved the aura and the sense of connectedness that I felt with my brethren as I learned about and admired the many facets of their culture.

Reports from my public health nursing superiors noted that I learned rapidly and had a creative outlook in problem solving. Those factors, my genuine love of people, and one particular supervisor who had unusual faith in me, soon led me to more responsible positions. I was placed in a federally-financed Chronic Disease Program, which, in time, became the forerunner of the first Home Care program established in the state. It was a program initiating referrals from hospitals to public health nursing agencies, (again emphasizing the whole person as they interacted with their environment.) Its focus was on rehabilitation nursing and to improve communication among professionals and agencies. The Health Department staff had great confidence in me, affording me opportunities for growth and further prepared me professionally by sponsoring my attainment of a Bachelor of Science Degree in Nursing at Boston University. While there, I was elected to Nursing's International Honor Society, Sigma Theta Tau. Recipients of this award were recognized for leadership potential as well as academic performance.

Upon completion of my degree, I returned to work with the Health Department as a supervisor. With further mentoring and affirmation of my abilities and potential, I was instrumental in preparing public health nursing agencies for the onset of Medicare and for the rigorous standards required by that program.

For the most part, my nursing career was marked by intensity of spirit, a warm comradery among my colleagues, a wide range of work experience within diverse groups and the resulting appreciation of my dedication and accomplishments from superiors.

I could never begin to imagine, nor could they believe, that in later years when my daughter was experiencing problems, that I could be treated abusively and disrespectfully by younger, less experienced and less mature "professionals" - professionals who, at that time, could see me only in the roles of a wife and a mother. Others were the "professionals" then. But, alas, they seemed to consider themselves "experts" on everything: on the upbringing of my daughter, my marital relationship, and indeed, on all else about me. There was little room for appreciation of my credibility, of the knowledge that I had about the appropriate up-bringing of my children, of my value system, of the abundance of my professional experience or of the knowledge that led me to the correct interpretations of my daughter's medical problems.

Professionalism, in itself, is hopefully something to be admired, but many of those that we met on our journey used their positions of power and authority to intimidate and control, shutting themselves off to the idea that there might perhaps be other knowledge, experience or other unknown factors that might supersede their authority.

So begins my story, an incredible journey to hell and back, as I walked with my "gifted" daughter through the indefinable anguish of the problems of hyperactivity, extreme allergic reactions, immune system deficiencies, and later pubertal, hormonal changes which gave rise to symptoms that some would call agoraphobia, chronic fatigue syndrome, and other neurochemical, neurobiological and neurohormonal "disorders". Together, we walked through many labels--and survived!

Today we each have just one label in which we all share. It is called---Creatively Gifted.

Read on.

The Story

From Nursing Career to Family Career
Mom's Prenatal Health as Relevant

My husband-to-be was sixteen years my senior. We had met on a blind date. He possessed a high degree of sensitivity, loyalty to family values and patriotism. In high school, he had been a star basketball player and had planned on going to college. Instead he had been drafted into World War II at the age of eighteen and was wounded in combat. He returned home from Germany, with a Purple Heart and many other medals, but severe wounds had left him with limited use of his right arm. During and after his recovery, he volunteered his time to youth groups and was soon offered part-time employment at the local Boys' Club. This led to full time work there and soon he became unit director, a position that he held when we met.

I was strongly impressed by the solid Italian heritage which he demonstrated, - a first generation American, his parents having immigrated to this country in pursuit of the "American Dream". They were proud to be counted among this country's citizens and even prouder of the three sons who had served valiantly during the war. Warmed and intrigued by his large family and their Italian traditions, I knew that this was to be the family that I wanted for my children's heritage. We married in grand style a year and a half later.

Soon after, I became pregnant, but miscarried. Three months later, I became pregnant again and was immediately placed on medication to prevent another miscarriage. I was not surprised at all this. For even though I had maintained a full, active and productive life, I had experienced many and varied metabolic problems for which I had been treated throughout the years. Moreover, my reproductive system had never functioned properly. At 21 years of age, I was told that a hysterectomy could be a reality in the near future and that I might never have children. Fortunately, the problem was able to be controlled by the removal of an ovarian cyst and follow-up treatments to correct the problem and to prevent recurrence of the extremely low blood counts which had become a "norm" for me.

A few years later, I took a routine glucose-tolerance test at a community screening clinic which I was staffing. I thought it a good idea to be screened since my father had just been diagnosed as having diabetes. I was still surprised, though, when I was called back for a second test and later told that I was "pre-diabetic" and I could expect to develop diabetes during adulthood.

Needless to say, my metabolic system had its flaws. But by now, during this pregnancy, most of my health problems seemed to be under control. The only medication that I was taking was a thyroid preparation, originally prescribed in nursing school for an "overweight" problem. I was told then that I would have to take it throughout my life for what they termed "hypothyroidism". That and the medication to prevent another miscarriage seemed to be all I needed at the time. All in all, that was pretty good, I thought.

Unfortunately, I smoked. Even though I knew of the gross dangers of smoking, I had taken it up about eight years earlier in an attempt to try to curtail a long-term weight problem. Ironically, it seemed to. The fact of my smoking almost one and a half packages of cigarettes per day did not seem to be of great concern to the obstetrician. (Unfortunately, we did not have all of the gruesome statistics about low birth weights and other fetal and childhood health problems that we now know to be related to smoking.)

I would categorize my chosen obstetrician as a strict one, quiet, but strict. Although we got along well he was always accusing me of "cheating" on my low sodium diet. He always found me to have fluid retention. To him there was no other reason for it except that I had to be "cheating", regardless of the truth that I was not. I was repeatedly discouraged at these accusations, knowing how carefully I was following that low-sodium diet. However, I viewed them as more of the same kinds of remarks that had often been made about my physical size, my allergies and other illnesses prior to the medical findings which gave them labels and reasons were discovered for them. (By this same reasoning, didn't it make a difference now, I thought, that my maternal grandmother had weighed 240+ pounds, and been diagnosed with Bright's disease before she died?) But for then, I could put up with the stringent low sodium diet and the false accusations of "cheating", for I was thrilled with the pregnancy!

At that time, the advanced diagnostic tests to detect subtle abnormalities and predict the sex of the fetus were not routinely done. And there seemed to be no need for me to have them. I had frequent routine check-ups, and there were no gross startling factors that warranted any specific non-routine care or investigation. Feeling good, I decided to continue work until late in the pregnancy--and I did. I was a nursing supervisor, earning a good salary and benefits, but nevertheless, I had planned to resign near the expected time of delivery to become a full-time wife and mother. This was not uncommon then and there was generally no scoffing by others at one's choice to do so.

As my due date drew near, I settled down to becoming an "at-home Mom". I became more and more attentive to the movements of the tiny child within me. To me, the graceful, smooth movements, that I was experiencing, indicated that this child was a ballerina or a swimmer. I could hardly contain my intense curiosity and excitement about the coming birth.

One evening in mid-October, the "ballerina" made a forceful statement of arrival. Shortly before midnight, while I walked through the kitchen, my membranes ruptured. Although I soon arrived at the hospital, I was in labor all night. For some reason, the staff nurses were not able to "feel" and time the contractions, so it was decided that I was not having them. I wondered, if like my metabolic flaws, I also had some strange tilt to my physical anatomy that could not indicate the strength of my contractions, for I was certain that indeed I was having them--and I was! And so it was, that without the advantage of natural child-birth class preparation, I experienced childbirth in its true, natural form. Medication, hastily given en route to the delivery room, did not take effect until well after the need for it. I was fully awake and able to appreciate that very first sight of my daughter. All I could say was, "she's beautiful"!

The Birth of an Angel

I will always remember that first sight of my daughter and of her first interaction with her father. I have seen many newborns at and following delivery, but I have never seen any infant so incredibly alert and wide-eyed as she gazed up at my husband. She vividly and impressively followed the sound of his voice. It was as though she was trying to absorb all of the earth's wonders in that one first glance. She weighed only 5 pounds, 8 ounces, but her relatively small size did not seem relative to the great amount of spirit that she seemed to possess. Every time that I saw her in the hospital, she was wide awake and *highly* alert.

I always knew when the nurses were bringing her down the corridor to be fed as she was *always* the one with the loud screech. I thought that she was always starved, for each time that I fed her formula, she gobbled it down as though she had not eaten in days. But just as quickly, she regurgitated it. Looking back, I see this now as just the beginning of a highly--motivated little person who just could not learn and do things fast enough to suit her.

Even after we went home, she continued to gulp and regurgitate. This went on for several months. Even though I continually brought this fact to the pediatrician's attention, he did not think anything of it. It persisted for several months in spite of a change to a different formula which was supposed to be more agreeable to most infants. That problem finally resolved itself when she began to drink from a cup at about nine months of age. She was petite, but, nevertheless she continued to gain weight and to be extremely active. She also continued to eat, drink and do everything as quickly as possible with an extremely visible high motivational level.

Our pediatrician was well-known and highly regarded in the medical arena, in the community and in the local hospital. He assured me that there was no cause for concern, that I was just a "nervous mother". Having been assured that there was no cause for further investigation, we just kept a bib on her most of the time. I did not worry and continued to enjoy this little "bundle of energy". I did *not*, however, like being termed a "nervous mother". I did not know then that was one of the first manifestations of the many problems that we would handle together and the beginning of new labels for me.

Karen, (as we named her), continued to be highly curious and highly motivated. She seemed constantly to want to do more than what her size and maturity enabled her to do and became highly frustrated when she did not succeed at attempts. When lying in the playpen, she cried when she was unable to sit. Once she accomplished that, she cried when she could not get herself up to a standing position, even though she was probably only about six months old. When she finally did that she was ready to walk with each of us holding a hand beside her but she did not become tired when we did and always cried when we stopped walking. She walked by herself at eight and a half months of age and from then on she was off and running. She played intently and responded playfully and vigorously.

She was a highly active child, demanding both supervision and soothing. She would become exhausted from the high levels of energy that she was expending and seemed unable to wind down. Every night I would sit with her in the rocking chair and sing to her before I placed her in the crib with quiet music from a mobile. In spite of my efforts, this did not always calm her. Many times I tried just leaving her in the crib crying for long periods of time, hoping that this would work. It did not. Sometimes, we resorted to walking her while talking softly to her before she would drop off to slumber. At times it took more than an hour of walking to get her to sleep. Although I was often exhausted, I was not alarmed. I knew many people who had experienced nights of floor-walking with so-called "colicky" infants. And besides, my husband was really good at that. He loved being with her and had more patience to do this at the end of the day than I did.

I was beginning to hear the term "hyperactive kids". How active did one have to be to be called "hyperactive"?, I thought. But certainly my daughter did not fit into this category. The pediatrician was aware of our problems, but he seemed to think that it was all just part of the "nervous mother syndrome". I had not asked nor had he offered medication and for that I was glad. I was not a person who sought medication as a "quick-fix" to solve problems. But, one day, when he finally realized that the problem had continued for several months and that lack of sleep was beginning to interfere with all of us, he registered surprise. He really didn't think that after so many months the "poor sleep habits" still existed. After some initial reluctance on my part, we finally agreed that at night, at least, she would be given medication containing a sedative. I did not like this, but it allowed us to get some much-needed "normal" sleep. For then it seemed that we had resolved whatever difficulties that had presented.

The visiting nurse soon came to see us as part of the State's Maternal Child Health Program. She had been assigned to visit us in conjunction with the same type of program in which I had visited infants as a public health nurse. The current focus was to visit the infants and mothers of those who were considered to be "at-risk". My medical history prior to and during my pregnancy, Karen's low birth weight, an Rh negative blood type and a few other "minor" statistics qualified us for an "at-risk" visit.)

It seemed that we did not have much to report in the way of existing problems. I did tell her about the formula regurgitation, enormous activity level and great difficulty that we had experienced in getting her to sleep. She empathized with me and we reviewed how I was handling the problems, but these, of themselves, were not outstanding ones that warranted continued nursing visits. In addition, the nurse who visited was from the very same agency where I first began my own professional work in public health nursing. She knew me from that relationship as a highly competent, intelligent and knowledgeable person. We talked about the foolishness of the "nervous mother syndrome", and left it that I would use my judgment and contact them if I were unable to handle the relatively few problems that my infant presented.

I loved the "Little One" whose presence had blessed our home (New Voices in American Poetry)[1] and attributed our few minor difficulties to normal child-rearing. I did not know that later on these subtle disturbances would change in nature and be called by some in their field as "neurobiological disorders". I did not know in my ignorance and disregard for labels, that in my attempt to bring up my daughter with love, that I would also turn these "hidden disabilities" into positive attributes, and that she would also be called "gifted".

Because of her vivaciousness, Karen soon acquired two nicknames. Her uncle called her "go-go" because it so aptly characterized her constant movement. Our next door neighbor called her "lungs" because she so often cried when she became frustrated at her inability to accomplish new things. My discussions with the pediatrician about her high energy level and other questions that I had about her intensively-driven behaviors were still met with the comment that I was a "nervous mother". I knew that I wasn't, felt very confident about my mothering and truly enjoyed my child. (In 1968 we did not have half of the working knowledge of the various subtle differences in the brain that we do now. Unless there was a serious, observable problem, one accepted individual differences as just that--differences. The term, "hyperactivity" was only being tossed about at that time as a "condition" that some kids had.) But could there possibly be something "wrong" with our child? We didn't think so. We were doing a fine job and loving our child, as was everyone in our large Italian family!

Karen continued to be highly active and highly sociable, as well. I was amazed at her first birthday, when she joyfully welcomed the little neighborhood children at the front door, seeming to already understand the nature of birthday parties.

But, by this time she had begun to suffer frequent bouts of bronchitis and what I now recall as the beginning of many allergic reactions. She ran a fever with every tooth she cut--sometimes as high as 103, and she still seemed to have trouble digesting milk. I would later learn that she lacked a sufficient amount of enzymes to digest milk products.

Although my days were busy with her, I was anxious to have more children. My chosen role at that time was motherhood. I became pregnant again when she was only nine months old. I was ready, and I believed she also would be ready to welcome a sibling in another nine months. She was walking, emulating the things I did around the house, and was always eager to help. I continued to be awed and overjoyed with her and eagerly awaited the arrival of my next child.

The Arrival of a Brother

I knew that I was carrying a boy. His movements were so much different - so firm, strong and swift. We hadn't even selected a female name, because of my confidence that this child would be a male. The pregnancy went well and without incident up until the time of delivery. Once again, before I even went to the hospital, I could feel the labor pains of contraction in my back. I called the obstetrician who was covering and he told me to get to the hospital immediately. I arrived, thinking for sure that I was in the advanced stages of labor, but was told that I would not be delivering for at least several hours. Again, they could not confirm my contractions. They didn't even think that they were strong enough to even warrant hospitalization yet! Therefore, they would call a resident to confirm that and request that I be discharged. I thought to myself, "if this is not labor, then what is"? I knew that at home, I was having contractions at three to four minutes apart with severe pain in my lower back. How I was going to be able to withstand further pain if this was not considered labor! But, when the resident examined me for discharge, he proclaimed excitedly that I was ready to deliver at that *very* moment! Once again, I had no medication. I remember yelling, as they wheeled me across the hall to the labor room,"we're not going to make it"! Amidst a frenzy, I was quickly given a "spinal" and rolled back just in time to deliver my baby in an appropriate position. Once again, inadvertently, I was able to watch and be an active part of the delivery proceedings. I was thrilled when they announced that my baby was indeed a boy!

Very much unlike my daughter, he was a big baby, with long fingers and toes - always asleep and always content. When the nurses brought him in for feeding, it always took prodding for me to waken him and to keep him awake while he drank his formula. And, unlike Karen, he did not "spit-up".

I had prepared Karen for his arrival in the best way that I could. Beside the infant dressing table, there was a small T.V. table. It contained an empty Vaseline jar, an empty can of baby powder and a few diapers on it. It was Karen's special table for her "baby". She loved changing her "baby", a cloth "Humpty" doll made by her godmother, and readily emulated me. I scheduled times during the day that she knew were for just her and I. Together, we participated in the various activities of "Sesame Street" walking "straight and tall" with baskets on our heads, and playing "tambourines". I helped her to strut on the, then-popular, "romper-stompers" and she greatly prided herself at this accomplishment. We never considered sending her to nursery school or day care. That form of child care was nowhere near as prevalent as it now and we had no need or desire for it. Karen mixed readily and easily with children of her own age and the new housing development that had been built next door to our home was increasingly providing an influx of young families with small children. She had the advantage of playing and learning with other children in the same types of activities as she would in any nursery school and, in addition she was always surrounded by a large loving family. That arrangement was very satisfactory to us.

Our family values continued to be directed toward enjoying the traditions of our large Italian family. I was happy as an "at-home Mom", and by now, my husband had become Executive Director of the local Boy's Club. He had a working knowledge of the tasks involved in running the Club and long-term generational relationships with many members of the community, whereas I had previous experience with the workings of community agencies and was familiar with the writing of "goal-setting" proposals. That made for a good team. I had always liked writing task objectives and proposals and had done well with administrative duties. I was able to use some of my professional experience and knowledge for the club's benefit, an endeavor which was not only was helpful to him, but it also provided an opportunity for me to use some of my former skills, updating and strengthening my resume. Personally, I was able to satisfy my need for intellectual tasks, while reaping the rewards of the challenges of child-rearing, cooking and homemaking. I did not feel undermined in any way as a "stay-at-home Mom" even while other moms were gradually resuming a place in the work force. I had the "best of both worlds", and planned eventually to prepare myself to return to the professional field, once the children were older. Right now things seemed to be going pretty well, all 'round.

My son was progressing in his own development and increasing in individuality. He was, in contrast to Karen, very quiet. Although one might be tempted to call him passive, such was not the case. He viewed everything with solemn intensity. He sat in his carriage for long periods of time, cooing at the leaves on the trees in the backyard, almost as if he were playing with them. When he was older and went out to play, he was still very quiet. Although he mingled with the other children readily, he did not join in the rough and tumble play. But it was apparent that he observed much as he sat on the grass, and that he enjoyed watching the other children run around him.

Karen loved her brother dearly and amply expressed that love. She talked to him and explained things to him. From the time of his birth, she truly was a "big sister" to him. When cards arrived for his first birthday, she showed them to him, telling him who had sent each one. Although only two and a half years old, she could recognize and accurately identify many things and still loved to help with everything. One day when she was dusting the windowsill in her bedroom, I remarked about all the dust that had accumulated, and she turned her little self to me and said, "Mommy, that's not dust, that's air pollution". It was about this time, that I began to regard and respect her mature awareness and seemingly intuitive knowledge.

I talked to my children often, explaining things to them as the occasion seemed to warrant, but very seldom did I sit down to teach them formally. I often read to them and remember so well those times when I sat between them on the living room sofa reading them stories, quietly marveling at their reactions. When my son was just two, I heard him repeating words, (ever so softly), from the stories that I was reading to them. As time went on, I realized that he was repeating and teaching himself words that were new for him. Shortly after this he was reciting the alphabet. He too, showed self-initiative for learning but in his own unique way.

I enrolled them in story hour at the local library. They loved the staff at the library and delighted in the way that they made books and stories come alive for them in song and play. But, I began to hear words of criticism from other parents that I was "pushing" my kids, even with a simple little thing as regular attendance at the Story Hours. This bothered me, but, as I evaluated the comments, I realized that I loved my children, was aware of their high quest for knowledge and was helping them to satisfy it.

Karen continued to show a marked ability to learn things readily. Not only did she demonstrate a keen alertness toward knowledge and people, but she also demonstrated other kinds of sensitivities. She seemed very sensitive to the environment and to a number of foods, as well. She ate well of mostly home--made foods, but excema, red facial blotches, boils from wet bathing suits, frequent fevers and severe constipation became more the norm than not. I had long-since stopped smoking, but I still believe that this earlier habit may have had a lot to do with her little developing system before I had realized it.

In spite of an increase in the kinds and numbers of subtle problems, she maintained her enormous, constant output of energy, always looking for things to do. She was always the first one on and the last one off the climbing bars, swings or whatever other challenging things she could find. I watched her keep up and surpass, even the little boys at these feats and decided that she needed some way to productively channel some of this energy. So, when she was only three, I enrolled her in a local dance studio.

I thought for sure that the little tap dances for three year olds would satisfy some of her needs and they seemed to. She absolutely loved them. But, even more surprising to me, was that she loved the ballet classes, (which accompanied them), even more. It was not long before she constantly twirled about the kitchen, hand positioned above her head, in a world of her own, a world which would later become so very much a part of her.

Music and dance were a big part of our house. It began with the little plastic pianos and my husband leading the parade of two through the kitchen with their wooden spoons, banging on pans. Through the years, it changed in form, but progressed. With the little guitar-playing that I remembered, I managed to play enough cords to sing songs in the summertime and at Christmas. My husband had played the harmonica years before with his brothers, and more than once, we enjoyed a "dusty" version of the old harmonica-playing at the children's birthday parties.

Although we had fun times, my husband was very busy with his job. Since he became Executive Director he worked even longer hours than previously and often put in many hours on weekends. The income from a human services agency position, even at the executive level, was very low, but it was our choice to live on one income. We both were conservative spenders and good money managers. He had experienced the financial hardships of the Great Depression and I had always been used to saving money before making purchases. In that, we held a commonality. We honored "old-time" family values of being able to do without things that were frivolous, showy, or not needed. From early on, we tried to teach these values and budgeting skills to the children. Even our vacations were economically sound. We loved nature and the simple pleasures of life. We went on short vacation trips to Storyland; to the mountains and lakes of New Hampshire; we took day trips to zoos, to museums, and to parks, to feed the ducks and ride carrousels. As the children got older, I took them to Pennsylvania Dutch Country, to Boston via train and on other educational and enjoyable ventures. In general, we tried to expose them to life's good things.

Each Sunday, when we were able, we packed up clothes, playpen, food and whatever else we would need for an enjoyable day and headed out to be with my parents. They were retired and lived on a lake at the end of a long dirt road. In all seasons, it was a vacation land in itself. There was fishing, boating, swimming, hiking, sledding, acorn-picking, - all of those wonderful things for young people to do on a Sunday afternoon. My parents enjoyed the children immensely and appreciated the many "odd jobs" that my husband did for them. My father always eagerly awaited the children's visits and loved being the "gatekeeper" of the large toychest in his bedroom, kept aside for the grandchildren's visits. My mother, still skilled in the craft works of her many years as a Girl Scout leader, made all sorts of things for the children... wooden racetracks, dolls clothes, hand-carved ducks, games and home-made fresh-baked "goodies".

There was much joy in our visits to the lakeside home, especially on the Fourth of July. On that day, we followed a tradition that I started there. Both sets of families gathered for a huge all-day get-together on the lake with a mixture of all ages and foods of all kinds. There was swimming in the beautiful lake, and rides in our old rowboat. Infants, teenagers and seniors enjoyed themselves and the spirit of "the 4th" was truly there! Later in the day, we brought the children to the local parade and the day's events culminated in a cook-out.

In like manner, my husband's family hosted the traditional

Christmas Eve celebrations. Every year, my sister-in-law, (the matriarch), prepared the traditional fish dinner for family members, extended family and many friends. Santa came and all members looked forward to this special occasion. We are all older now and a new generation has sprung up, but this beautiful Christmas tradition continues.

Birthdays, holidays and other traditional ceremonial celebrations were big occasions for us, particularly in view of the large Italian family that always surrounded us. My husband was very close to the traditional way of life that he had always known, and observed the cultural ways of life in strict Italian fashion. He was emotional and quick of temper, but loved his family and his children dearly. Whenever he could, he sat with them in the evening after supper, building blocks or engaging in some other playtime activity. Everything had to be family centered. In short, we were no different from the many large ethnic families who retained their culture and identity. The unfortunate part was that, as traditional family values and two-parent families became less prevalent, our lifestyles and value systems would soon become the minority ones in a rapidly changing society.

"Real" Problems of a "Real" Family

Our family was known as a "nice" family and we were even once characterized as an "All-American Family". But we experienced the growing pains that all persons and relationships do as they participate in the daily living/learning process. Families do not exist without disagreements, without differences, without varying abilities and interests. The difficulties lie in how to make them be positive for all concerned, particularly when there are growing children within a family.

And as is so often in families and relationships, the same characteristics that I so dearly valued in our family were often the very ones that often got in the way. Although my husband was very content to have his family be his sole source of diversion and relaxation, I was eager for outside stimulation and the comradery that comes from participation in enjoyable social activities. I needed occasionally to remove myself from the constant demands of family, however much I enjoyed them. I needed to recreate myself. After all, my sense of adventure and love of fun did not stop just because I became a parent. I wanted for us to get away for a week-end by ourselves, without the children, but this would not do. Getting baby-sitters were problematic unless the sitter was a family member or close friend and then it should not be for "over-nighters". I admired his traditional ways but my desire for freedom for myself and the children began to clash. I had a need to recreate myself for the benefit of myself and my family. This drive had nothing at all to do with any part of the feminist movement that was now beginning to happen; I just wanted to use my God-given ability and skills. I needed other challenges.

There often were outbursts about our different needs and

values. His long hours of work, including evenings and some weekends, further limited my lack of outside diversion and challenge. I believe this sense of isolation and lack of intellectual interactions, often contributed to a loss of my patience as I tried to cope with Karen's many and varied needs.

Throughout, I was not only often frustrated, but puzzled. Karen's physical problems still remained very prominent. Her intensity of spirit, even in play and on joyful occasions was exuberant, buy just as easily, seemed to result in total physical burn-out. The symptoms that I had noted from her birth seemed to be increasing in kind and severity. It wasn't long before I realized that Karen had a high degree and wide range of sensitivities - to foods, to people's feelings, to the environment, to just about everything. Each time she received childhood immunizations, she developed a local reaction as well as a fever as high as 103. Although the pediatrician reduced follow-up injections to only half the usual dose, (given in two intervals), she still reacted. She often became extremely constipated for long periods; when she played outside in hot or humid weather, she became fatigued to the point where she often became very red-faced and even glassy-eyed. At these times, I slowed her pace and sometimes even brought her inside to get the benefit of air-conditioning, depending on the severity of her condition. When she became completely exhausted, she would lose her voice as well. She still had difficulty in quieting to get to sleep, in spite of her exhaustion, but once asleep, daytime naps often lasted three hours. She easily developed red blotches on her face and developed excema; her skin was very sensitive, often becoming red when barely touched; she sometimes developed tiny bubbles when she would easily scratch an itch. Each time she cut a tooth, she became feverish and highly irritable. One dentist (having noted this in other children) told me that she was probably allergic to her own saliva. (I later wondered, but only to myself, was she exhibiting some sort of early auto-immune problems?)

Not only were these symptoms difficult to handle at home, but they also began to pose problems in regard to her socializing. Every time she went to a party, she would become exhausted from the excitement and activity and react badly to the various punches and food colorings. I began to be conflicted as to how best to handle the problems. I hated to restrict her freedom of spirit in any way and wondered, "how can I allow her spirit to grow and flourish when there always seemed to be these contra-indications"? Karen truly embodied the saying, "the spirit is willing, but the flesh is weak". She wanted to do so much, was so sociable and outgoing, yet became so tired.

I loved it when she was given a chance to go to Grammy and Grampy's house each summer. It was a chance for her to be free and to romp in the nature that she loved. But, even in spite of Grammy's love and best efforts, she would have much difficulty. Her high energetic output, an ever so slight change in diet, the humidity of the season, the dampness of the cottage and surrounding woods somehow always seemed to interfere with the benefits of the pleasure. She loved it there, but even with the best of care, in a few days there would always be the slightly elevated temperature, the boils from a wet bathing suit, the deep purple under her eyes and a completely exhausted little girl.

Nothing that I had reported to the pediatrician seemed to strike him as being a problem or cause for investigation. My son, eighteen months younger than Karen, seemed to be doing very well. He was developing differently but was in no way problematic. Friends and associates said to me, "girls are easy, just wait until your son begins". But he never had those kinds of problems. He never showed the symptoms or sensitivities that Karen had. (I often think of that dentist whom she visited at age three, and believe that he, alone, had the most accurate knowledge.) There was something wrong with Karen's immune system. Yet, even if the symptoms were diagnosed or termed as such, could we really have done anything differently from what we did? I doubt it. We just lived and managed in the best way that we were able and prepared for the next big event...the beginning of school.

Ready for School

Karen and I had both looked forward to her first day at school. It was the time she had long awaited. Each September, she had watched the neighborhood children as they returned to school after summer holidays, eagerly awaiting the time when she could join them. Many mornings she sat in front of the big window in the living room with her toy thermos and lunch box, watching the children pass by on their way to school.

The numbers of children living in the plat were rapidly growing and consisted of a wide variety of ethnic and income groups and she was friendly with all of them. We were happy that she had an opportunity so young to mingle with such a wide variety of people--and she had loved playing with all of them. But most had started school before her.

Now it was finally her turn. She had been so eager to attend school and she delighted in her first days at this new adventure. She welcomed it and loved it as she had all of the rest of the things in her life. She mixed readily with the other students and was a source of delight to her teachers. Her days in kindergarten were pleasant and successful ones. Attending school on a half-day basis was manageable. She was able to have her afternoon naptimes and maintain a half-way decent health pattern.

But first grade was different. They were all-day sessions and the beginning of a more formalized pattern of socialization. She was a popular little girl and was invited to many birthday parties. She excelled in her schoolwork and was always picked for activities that required competence and skill, such as the long parts in school plays, etc. In one way, she thrived there but various physical symptoms continued. The hoarseness became increasingly noticeable, particularly during humid times or in the winter dampness. In spite of her efforts and enthusiasm, when it came time for the rehearsal for the Christmas play, she had no voice, at all. Once again, the doctor had brushed it off with "put her in for a nap when she gets home". But this time, I would not "brush it off". I had other support. I had already spoken with her first grade teacher about the problems with which we had dealt. Thank God, that she had a warm spot for this "bright little girl" . She, too, was concerned about what was causing her to lose her voice.

She agreed that what she had observed was sufficient cause for concern and agreed to give input to the pediatrician, if needed. Observations would be coming from yet another perspective and from a peer professional. Maybe that's was it was that was needed for him to heed my concerns. Apparently it was, sad to say! For when I called to inform him that the teacher was willing to come to the visit with me, he finally thought that he should investigate further. He referred Karen to an ear, nose and throat specialist.

[I think it is important to keep in mind that Karen's symptomatology had so far consisted of a wide range of vague, non-specific, non-patterned, (or so it seemed), problems- each one minor- but together making for difficult living. If we had sought specialists for all of Karen's complaints and symptoms, we probably would have been on a diagnostic treadmill for months or years and still may have had no conclusive evidence of anything tangible. Some friends and acquaintances had already indicated that Karen was experiencing things that happen to people everyday, so why should I be concerned. That may be so, but what was to happen later does *not* happen to people every day.]

The specialist found that Karen's larynx was swollen, probably due to allergies, but he wanted to rule out that she might have a tumor on her larynx. He decided to do a tonsillectomy and, at the same time, conduct an exploration. The conclusive evidence was that allergies had caused her larynx to swell and that it would continue to happen. That was one way that she manifested allergic reactions, but he also wanted to follow-up with an investigation for food allergies.

A series of tests was conducted which revealed that she was allergic to many foods and also had environmental allergies. These, he thought, were partly responsible for the deep face--flushing, constipation and exhaustion she experienced when outside at play. At last, I felt we were beginning to get some answers.

The period of testing was long and difficult for all of us. The process was to eliminate various foods one at a time over a period of several days. A few specific foods could be eaten at a time and were to be eaten for a three-day period only. During that time we were supposed to monitor her for reactions.

During the testing, Karen was invited to several birthday parties. Even well-meaning friends and relatives just could not understand why it was necessary to keep her from "cheating". Normally, I would have been content to let her "cheat". Birthday parties are so important for a very outgoing little girl. But "cheating" did not get the testing process completed. And that's the way it was. I did my best and finally completed the three day food-free cycles for *each* food being tested, and reported my observations to the allergist. After several weeks, we had the results and a dietary regime which was thought would keep her symptoms at a minimum. She could not eat white flour, eggs, milk, cinnamon, wheat of any sort and many other foods which comprise the "typical American" diet. I was concerned as to how best to handle this with such a young child just beginning formal socialization. She was only six years old. She could understand that there were basic reasons why she had to eat certain foods, but as bright as she was, she did not understand the implications of it all. I believe that it was during this period that I had an "inkling" that at some point in Karen's life, she would have to face the decisions of maintaining dietary and environmental restrictions or having freedom of play, freedom of food choices and bedtimes and with that, the untoward consequences that might incur. The teen years would be the test of that. For now she was still only in first grade. For now she needed our guidance, support and parenting.

I tried to find different ways of setting up mealtimes for her that would make it easier for her to eat her non-appetizing food. When we were going to have something that she could not eat, I set up a special place in the den just for her with the T.V. (a treat for her, as watching T.V. was not a dinnertime activity in our home). Preparing her meals challenged my intelligence and creativity. I remember making doughnuts with oat flour, egg substitute and powdered cream. Jelly was one of the few things that she could safely have. I questioned what *quality* her nutritional level would be with this new diet as I was aware of the high and varied nutritional value of the foods that were being eliminated. I recall saying to the allergist, "It seems as though all she can take to school for lunch are jelly sandwiches". He replied, "If that's all she can have, then that's all she can have." I was not content with this, realizing that a steady diet of only those allowed foods would eventually cause other problems. And so I began my own personal research project into the value of nutrients, allergies, auto-immune disorders and their inter-relatedness.

I did not foresee, that in later years, these same problems, complicated by the hormonal changes of puberty would be exacerbated and change in nature and that my "parenting" would be blamed for the resulting symptoms. But, Thank God, I was doing my research.

Nutrients, Allergies and "Stress-related Conditions"

I sought information from many sources and learned much about natural foods, enzyme systems and what nutrients served what purpose. With my background knowledge of anatomy, physiology and experiential education that I had gained through many years of public health nursing, I was able to make a lot of sense out of what I was reading. I read about the functions of various vitamins and in what foods they could be found. I discovered that the"B" vitamins were very important and had a key role in enzyme-making. I began to research nutritional deficiencies as they related to diseases.

I began to ask questions about our family history and looked at the answers in a new light. I researched diseases and dietary habits. I discovered that my grandfather, whom I am told never ate vegetables, died while in his 50's with the cause of death determined to be Pellagra, a condition associated with a lack of nicotinic acid or a Niacin (B vitamin) deficiency and thought to be a very rare occurrence at the time.[2] I continuously read whatever I could find on research currently being done about these vitamins, being especially interested in the importance of vitamin B6 in the function of metabolism. I wondered if these B vitamins (or the lack of them), had anything to do with me and my previous metabolic problems. I learned more about enzyme deficiencies and slowly and cautiously began to take various vitamin, mineral and amino acid supplements myself. I wondered if a hereditary lack of nicotinic acid had anything to do with my reason for smoking and the fact that (although it had undesired side effects, it somehow seemed to relate to the control of my weight--in some, as yet unknown, physiological, way. I thought that these were good questions. They are ones on which comments are being made today by researchers in those fields.

I began to give my daughter B vitamins, only recently fully realizing the consequences that deficiencies of them can cause and their effects on the human systems, including that of the neurotransmitter system. I also became aware that deficiencies result, not only from the amounts ingested, but also by one's individual requirement for a particular vitamin and their body's ability to utilize it. In those things we all differ.

I paid close attention to what I was observing in my daughter as I started her on this new regime. I marveled at how my own health was improving since I started on a vitamin/mineral supplementation program.

Meanwhile, my cooking abilities were tremendously challenged. I bought food supplies from health food stores and found all sorts of interesting combinations that were tasty and high in nutrients. The Sunday trip to Grammy's now included a stop at a health food store to browse and select good, wholesome snacks. And this was still only in the 1970's! This was not any "New Age", but just plain common sense---a factor too often left behind in favor of new discoveries, new technologies and "wonder" drugs.

My husband had some trouble understanding and accepting the idea that various foods could cause so many problems. Seldom a meat eater, he had been brought up on a diet of natural and home-cooked foods -- greens, lentils, fish, pasta, dandelions,- and he was very healthy! He really did not seem to understand what all the fuss was about as related to foods and did not seem to share in my belief that nutrients or the lack of them could really contribute to such a wide variety of problems. I, on the other hand was gaining increasing knowledge and, therefore, acceptance that this was, indeed related to Karen's problems and also to alleviation of them.

Like most fathers of young children, he did not want to

come home to problems after many hours of work. He wanted to spend that time enjoying his children. He was not with them as much as I, and, naturally, was not able to observe all that I did. But, he became frustrated when my daughter would get exhausted to the point where we would have to leave family gatherings sooner than others, or when I insisted that she should not eat certain foods. He did not like the necessary pacing of activities, the fact that naps needed to be so strictly scheduled into our vacations, or the restrictions on foods, which interfered with his buying dairy treats. His very idea of "family" consisted of the food and festivities of the large family gatherings, of the children helping him rake grass and leaves, of the seemingly simple pleasure of buying them ice-cream cones. Yet, foods like ice cream were not allowed and both children had been started on desensitizing "shots" because of environmental allergies to pollens, trees, grasses and molds. This certainly was not compatible with helping with yard work and certainly were not helpful at all to our relationship. At times, he just seemed unable to fathom or accept that some "outside factor" should interfere with the joys of his childhood that he wanted to pass on to his children, but all in all, he tried to "go along" with it for Karen's sake.

"Branching Out"

We continued on with life in as normal a way as possible. I had worked out a reasonable diet for Karen. She could have beef, veal, lamb (roasted or broiled) and broiled chicken if all the fat was removed.

- Cereals were to be those of barley, rice, corn, oatmeal, tapioca, and rye meal.
- Flours were of rice, cornstarch, rye flour, oat flour, soy flour, pure buckwheat and starch flour. Bread was to be made from those flours - no white flour.
- She could have beets, beet-greens, spinach, carrots, sweet potatoes, and squash, bananas, pears, apples, grapes, apricots, peaches, plums, cranberries, raisins, and Juices were to be from apples, grapes, and prunes and desserts could be made from the above.
- Tea, coffee, cream substitutes were allowed. (Lactose-reduced milks were not yet available to us, so she was given a calcium substitute for that missing from lack of dairy products.)
- Jams and jellies could be made from the allowed fruits and, surprisingly to me, she could have salt, white sugar, brown sugar, maple sugar.
- Fats were to be of salt-free margarine and olive oil.

To anyone familiar with a therapeutic diet of any sort, this may not look too bad. I agree. But, for a six year old, just beginning to have lunches away from home with her peers, and always eager to try new things, it was trouble, in disguise. Although she did very well, it was sometimes not easy to refuse foods and, unfortunately, few seemed to realize the consequences of her cheating. "Food allergies" held different meanings for people. The various differential categories of allergies, food sensitivities and food intolerances were not clearly defined for the general public yet and she was often urged to "cheat". She and I were continually met with responses like, "we all have allergies."

It was hard enough trying to get her to understand, accept, and work with her condition at this young age, and comments like these especially from adults, made it even more difficult. I soon became tired of explaining that for her to "cheat" was to have one or more of a variety of reactions which could include an immediate rash, perhaps a fever, stomach aches, swelling of the face, frequent, long-lasting colds often leading to bronchitis, and usually resulted in school absences. Missing school was one of the most difficult things for her, as she loved school dearly. At times like these I could not help but think "what or where is the 'supposed' maturity of these adults"?

I did not tell them, as I wanted to, that they were trying to satisfy their own needs at the expense of helping a child to learn to handle the problems that life had handed her. I did not tell them how difficult it had been for her to have given up all of her stuffed animals, which had been part of the treatment to reduce allergic responses. (We had to rid her room of all possible dust collectors and things not made of hypo-allergenic substances. This category not only included the curtains continuing fiberglass material, that we found she was sensitive to, but practically all of her stuffed animals.) I do not think that these people truly understood the hardships that Karen was facing in her young life-- subtle, but because they were subtle and not obviously problematic is perhaps one of the reasons why they were difficult.

I soon learned that she and I together, had to be "bigger" than all the rest...that we had to try harder to understand others because it seemed that many would not care or want to understand *her* needs. How prophetic I was! And how glad I am that I had begun to prepare myself for this unfortunate fact so early in her young life.

Karen seemed to be sensitive to so many things. One day, she stood close beside me while I was changing her brother, readying to visit my parents. Her little head was leaning against me. As I looked down, I noticed that her face was very reddened and beginning to swell, so much so that we decided we'd better take her to an emergency room as we watched the swelling worsen. En route, the condition subsided. Later we found that the causative agent was the acrylic in the dress that I had been wearing!

It was very difficult, for her to be having such reactions and to continue to hear such comments as "we all have allergies". Even Karen's father, who had witnessed her reactions and felt very bad about her condition, often had difficulty understanding it all. Meanwhile, I was learning a lot about food sensitivities and the reality of "multiple chemical sensitivities", a name which I had not yet heard, but it was living with us.

We continued our struggle to live within the restrictions, while still trying to provide the things that would meet Karen's needs for socialization, and intellectual and recreational diversity.

In the summer following first grade, I found a summer camp offering attendance on a half-day basis, if desired. It was a private camp with a staff that was able to give individual attention to Karen's dietary and other needs. We gave it a try and it worked out well. Since it was only a half-day, she was able to come home to the air-conditioning unit, (which by now had been prescribed by the allergist), and to nap, if needed. I brought her each day to the bus stop where she went off with her snacks and camping gear for the morning. She was thrilled with the varied programs of crafts, games, live animals...and even ballet. I was pleased with the confidence she had demonstrated at going to this new place alone, so young, but, knowing her, was not surprised. She really wanted to go and I was satisfied that there was conscientious staff there who would help her individual needs. They did. She benefitted and flourished.

A New Pediatrician - A New Perspective "Normal" for Awhile

A new pediatrician's office opened a few blocks from our home. I was very willing to seek him out, even though he was new in town and did not have the well-known hospital affiliation as our current pediatrician. He was an easy going man with young children of his own...and they had allergies! He could appreciate everything that I was saying. He understood about frequent ear infections, the colds that did not clear up in the usual five days or so, the frequent deep purple coloring under my daughter's eyes and her extreme fatigue. In short, he was *knowledgeable* about her condition and presenting problems. He was truly interested and gave sound, practical advice and emotional support. I felt lucky, indeed!

He worked with the allergist to coordinate a program for Karen in which her symptoms were kept at a minimum. She and her brother flourished. I took them swimming regularly; they learned to take pictures, to do crafts. They made their own birthday and Christmas presents. They played outside with the other children in the neighborhood, (who, with the addition of the town houses next to our home, now numbered 27,) and took part in the activities held at the local Boys' Club. They were involved in so many things both in and outside of school. And the best part of all was that I seemed to have a routine for Karen which enabled her to do all this without serious consequences.

She remained on all natural foods, yogurt every day, and no additives or preservatives. She had a program of allergy shots and adhered to a schedule which included naps, early bed-times and special precautions in hot weather. She was doing well. We had finally found the proper care for her and a program to control her allergic sensitivities!

My son was also doing quite well. He had begun to develop firmly in his own way. He was quiet, but seemed to observe and absorb everything around him. He loved to sing and I loved to hear his strong child's voice as my daughter, he and I sang together. His sense of humor was a rare one. In short, he was a joy. (New Voices in American Poetry, 1989).[3] He taught me early in his young life to respect his individuality. In fact, it was through him, I think, that I learned my first lessons about sharing control with children. I admit that I was not quite sure at first for he was only two years old. (I am not speaking about the much-talked about stubbornness of the "terrible two's"), but rather a demonstration of individuality that almost seemed to be inherent within him. One day, I had insisted he could not go to the park with us unless he ate his carrots. He did not do so and did not go. I was proud of myself - as difficult as it had been for me to leave him behind, - I thought I was making a good start in teaching him that there would be natural consequences to his actions. I was the one who was fooled, however, when I later learned that every time he ate any appreciable amount of carrots, his face became orange-colored. There was some physiological aspect to how he did or did not respond to carotene. He had the right idea about what constituted "natural" consequences. I was to learn even more from him.

He was only about four years old when he, Karen, and I were making cookies, rolling out dough and decorating. As I showed them how to do it, he said to me very confidently and nonchalantly, "Mommy, I have a way." His calm, self-confidence about this amazed me and I stood back and watched him perform in his "way". I continued to be mindful of his early stance for individuality but it was not always easy to know when the ability accompanied the readiness. I guess that what happens to most people in adolescence was happening to him at about four years of age. And, much to the astonishment of many adults, I continued to let him develop that individuality in most instances throughout his growing up years.

We decided to enroll him in a small, informal nursery school. He was one of the last of the very large group of neighborhood children who had not yet gone off to school and I thought that he, in particular would stand to benefit from this type of socialization. The teacher of the nursery school was very impressed by his abilities. One day when she brought him home, she said to me, "Someday this boy will be able to move mountains if he has the confidence". I watched him grow and tried to give him that confidence.

I was, therefore, more prepared for his accomplishments when he went to kindergarten than I had been with Karen, and it didn't take long for those accomplishments to show. His kindergarten teacher contacted me very quickly to ask if I wanted to enroll him in a new program where he would go to first grade for reading. They had noticed that he was very advanced in reading readiness skills and did not wish to restrict him, she said. But this also meant that he would not be with the same classmates with whom he was just beginning to develop friendships. I was very mindful of my son's needs to grow in social interactions and of his reticence in interactions. To me his developing socialization was just as important as his developing reading ability. Keeping in mind his strong sense of his "self" and his "way", I told the teacher that I would discuss it with him. She seemed very surprised that such a matter would be discussed with a five year old. To my way of thinking, it was never too early to be open to the wisdom that begins in childhood. So he and I talked and he agreed to try the advanced reading program. Special provisions were made so that he could remain with his age-level comrades and still benefit from the reading program. There was one happy teacher and one satisfied parent when his expressed needs played a major role in this important negotiation.

The New Label "Giftedness"

It was increasingly being noted and affirmed that both children had talents and abilities, which were being recognized by others as "special". My daughter continued to exhibit high levels of confidence, sociability, and, above all, creativity. She was still highly curious, eager to see and do everything. My son, having received the benefits of the advanced reading program was showing potential in other areas as well. When his I.Q. was tested in first grade, I was informed it was about that of about a ten year old.

Both children earned school awards as results of their efforts, but self-appointed seers in the guise of other parents told me that I was "pushing them". Nothing could have been further from the truth. I merely provided opportunities and fostered their growth with gentle, loving support. But, being the conscientious person that I am, the remarks bothered me and, on at least one occasion, interfered with the easy flow of interaction that I usually had with my children.

One day just before Christmas, when my son was seven years old, he was unusually busy and quiet in his bedroom. Upon investigation, I found him making little numerical figures on scraps of gray cardboard, that he had torn into little rectangular pieces. On his desk, was a pencil that belonged to Karen which had a device that produced equations when rotated. I did not know what he was doing, but mindful of the recent remarks aimed at me, I said, "Ray, you shouldn't be doing that, this is your playtime. Why don't you color or something? (How stupid of me!) Very solemnly, he told me that he was making "flash cards" for his sister who was studying equations in school.

On Christmas morning, when Karen unwrapped the gift from her brother, there they were...perfectly numbered equations, lovingly pencilled on grey cardboard paper, torn into the shape of little cards. I tried never again to listen to the judgmental statements of other parents (and I still have those precious equation cards as a reminder of his love).

Happenings such as this made me carefully evaluate myself as a parent, just as I had always evaluated myself as a person. I knew myself fairly well, I thought. I had a life which I pursued outside of my children's lives and always tried to keep a sense of balance. I enjoyed activities in which I could socialize in a variety of ways with people of diverse interests. I found other women who shared my interests and went with them to such activities as line-dancing, the local repertory theatre and other cultural events. I swam regularly and bicycled on the streets around the neighborhood before it became popular for adults to do so. (I even remember a little girl excitedly pointing to me when I was bike-riding, saying loudly to the adult accompanying her, "Mommy, look at the lady"!) In general, I pursued, what I thought was a very balanced, open and experiential approach to life and its problems and I know that my children received some of the benefits of that lifestyle. I did not know that I was then practicing many aspects of what would later be recommended by many as "holistic" lifestyle. I just lived in the best way that I could.

Both children continued to enjoy school and the teachers enjoyed them. But it was truly a surprise when Karen's third grade teacher informed me that she was being considered for the gifted program in the city. I knew nothing about this program nor had I ever thought about any special class placement. Her teacher asked me if I had not noticed that my daughter was unusually bright and sociable. I really had not thought much at that time about her being unusual, as I had when she was younger, because her brother seemed to be equally as bright. He demonstrated his abilities differently, but was still bright.

Therefore, I thought, they must both be "normal". But the teacher informed me that Karen was exceptional, both socially and academically, and for those reasons she was recommending that my daughter be in the relatively new program for "gifted".

I was sent information about the screening and selection processes. Considerations were to be based on the child's performance since entering school. Abilities and attributes were evaluated on objective and subjective findings and included:
Kindergarten Metropolitan Readiness (those who received scores of "A" and a review of percentile ratings); Group Intelligence Test; Otis-Lennon Mental Ability Test (those receiving scores of 120 I.Q. and above, their percentile and Stanine; Gates-Mac Ginitie Reading Tests (those receiving scores two grades above grade level and their percentiles.)

That data was accompanied with recommendations from the school personnel and a form completed by the school principal and was then forwarded to the screening committee. They decided which students should be further tested with an individual intelligence test. All data was then reviewed anonymously by the Screening Committee, a number being the only identification. (No children's names accompanied the data.) From this, the final selection included those with the following qualifying criteria:
Group intelligence test results of I.Q. 120 and over and an I.Q. of 130 or above on an individual intelligence test; Consistent superior daily achievement in all academic subjects; Reading test scores in the 90th percentile range; Opinion of parents; Opinion of principals and teachers; Opinion of the child himself.
They were also evaluated on other traits;
Socially- relationships, etc.; Emotionally - maturity, stability, etc.; Area of special talent; Area of unusual interest; Personal health and physical history.

I was notified that Karen had met all of the criteria. I was thrilled at the knowledge, but this prompted many thoughts. I wanted her to have the most expansive opportunity in which to promote her potential, but at the same time, I did not want her to be in a situation where there would be pressure to perform and possibility of labels to follow. I had no idea of what the implications of this might be.

I made an appointment to speak with the principal of the school where the program would be held. I asked a lot of questions. I questioned the program itself, what it would mean in terms of placement with other children; in terms of pressure to perform academically; and what kind of follow-up there would be in junior high, (the program only continued through sixth grade). Those were good questions and I think that I was given the best answers that were available at that time.

I was told that she would have to leave her regular school district and be bussed daily to another district. There she would attend grades four through six and participate in the gifted program. Upon completion of grade six, she would return to the district where she would attend regular junior high. There would be no extension of the program, but she would automatically be placed in the accelerated program. Supposedly, at that time she would reunite with the children with whom she was now attending elementary school. It would be that easy. She would receive the benefit of the gifted program with "no strings attached".

That was the part that I should have looked at more carefully. She would be the only one sent to the program from her school that year. That part seemed to be O.K., but it also meant that she would probably also be in isolation when she returned to the district for junior high where peer group formation was so important and was already occurring on the elementary level. Without that knowledge, (gained in retrospect), I discussed with Karen all the information that I had received. I told her about the psychological tests that would have to be done and that she would go to another school district, which would mean leaving her friends.

Always eager for new experiences, she said somewhat

excitedly, "I'll make new friends, Mommy." I was proud of her confidence and felt quite certain that she would do well in her new setting. My husband agreed.

And so she was one of fourteen students city-wide, who had been tested and chosen by the school department and found to be an apt candidate for the fairly new program for gifted students. She would begin in the fall.

Ties to Tradition

Our family began to grow into the kind of family that I was happy with. For the most part, we remained steeped in tradition and members of the Catholic faith. So when the time of Karen's first communion arrived, we looked forward to it as a special occasion. The little Italian church that we attended was trying something new that year. We were given an option of having the traditional setting and ceremony for the reception of her first communion, or of planning a ceremony which differed from the traditional.

My brother, was a priest and professor, now in Europe. He was a biblical scholar and had been teaching at a seminary there for several years. When he returned to the states, he always spent time with the children and they were very close to him. We decided that we would have a small at-home ceremony on Christmas Eve in which my brother would say Mass and Karen would make her First Communion in the presence of family members. Seeking to connect the symbolism of our ceremony to the larger church community, Karen and I went quietly to the little Italian church that Christmas Eve afternoon. We placed a small, white poinsettia on the altar which remained there throughout the Christmas Season with all of the larger red ones. To me it was symbolic, even then. I knew she was a special little person.

The parish church that we attended was one in which my husband's family had many ties. A new building had just recently replaced the older structure which his father, a construction worker, had helped to build. It stood directly across the street from his family's home, still habitated by his sister and her family. It was where my husband had been an altar boy and later on a "collector," and where we had attended services since our marriage. I had taught singing and religious instruction (CCD) there. The instructions that I gave were a little different from the traditional way of teaching CCD then. My "lessons" reflected my beliefs in a spiritual scope - our connectedness to each other, to our environment and to the universe. My group's project for the year was to find and celebrate God's gift of living things. And so at Thanksgiving time, my third grade class brought their plants, pumpkins and other items to church to display their gratitude for earth's bounty. The Italian ethnic neighborhood for which that church was established, was slowly but increasingly, becoming more open to new traditions as new generations sprang up, and I am happy that so long age, I introduced some natural concepts of "church" as I know it, into the established church.

My children grew up in that Catholic tradition, which they revered. We appreciated the beauty of the songs and stories and they were active participants. Their young minds absorbed the stories about myths and angels and we had many expressions of those traditions. Christmas, of course, was very special, and my children were always very full of its symbolic spirit. But one particular Christmas, when the children were in first and third grades, respectively, there were many quiet whispers going on around me. At this time of year suspecting parents were not to question their offspring too much. But one day, I was told there was a "secret". I was not to know about it, but they needed me. They needed me to pick up Auntie Muffie, and take them all on a "secret errand". I found out later that they needed to buy some figurines for a manger--not just any figurines, but special ones. They had their own money, but they needed Auntie Muffie's help. I paid little attention on that December afternoon, keeping my end of the bargain and not asking questions. How glad that I did! On Christmas, there was a beautiful hand-made wooden manger from Ray, my son. It had been made from scraps of wood; it had a roughly-hewn, slanted roof, with an exactness that could meet any architectural expectations. Tiny trees lined the sides of the wooden floor, and the small figurines were nestled in the rough straw. There was a note beside the masterpiece, in my seven-year-old son's tiny handwriting. It said "This is the best I can do." The cherished gift lasts. It is irreplaceable and precious and has remained as sturdy and as beautiful as it was on that Christmas morning some seventeen years ago. Each Christmas it is carefully set up on the living room table with an additional package of love from Karen beneath it.

My son's talents and sensitivities were beginning to manifest themselves in so many ways. He loved music, art and the beauty of nature. He constantly drew - everything in sight, it seemed -- birds, animals, baseball, and his favorite cartoon characters and always so detailed and amazing for a seven year old! While Karen was winning recognition for her poems and stories, he was gaining recognition for the talents of his hands.

His early ability for fine finger dexterity remained obvious. Because of that and his continued his love of music, we decided to consider an offer of accordion lessons that was brought to us by a door-to-door salesman. Sure enough, he learned quickly and well and it was not long before he was walking around the house or the yard, with a big accordion strapped to his middle, playing the ever-popular accordion classic "Lady of Spain". He played lively and well. But it was not long before he told us that he did not need to practice. Although he liked to play and learned new songs quickly, he found practice boring. I really did not know whether to let him continue his lessons or just to let him continue on his own terms, without practice. He and I negotiated an agreement. He would spend some of his allowance on the lessons and on any registration fees for competitions that he wanted to enter. In that way, he would share in some of the responsibility. He readily agreed to that and his instructor, thoroughly impressed by his playing and the ease with which he learned, entered him in competitions both in and out of state. He practiced very little and each time he entered, he came home with a high placing trophy!

Money for these extra cultural activities was provided by a joint effort of the three of us. I was doing at-home telemarketing to pay for the "extras" while the children contributed to their own expenses by putting a small portion of their own meager allowances toward their music and dance lessons. Karen also had a fund in which she saved to help pay the cost of her dance costumes for the recitals. In this way, they learned the "old--fashioned" value of money.

I believed that I was doing "right" by teaching them the value of work and the earning of money in exchange for energy, even though their allowances were small and did not compare in any way to the money most of the neighborhood kids seemed to have. Most of the other kids could be found running to the ice-cream man's truck, seeming to be able to buy whatever kind of specialty ice cream they wanted at any time of day. My standards were different. And no, my children did not always like this. Do not all children want what everyone else seems to have, whether or not it seems, right, sensible or affordable? Still, I raised them according to my value system, in spite of a society that was rapidly changing to meet every whim and demands of children. Karen's diet was another reason why trips to the ice-cream truck were limited, but I tried to compensate. I made frozen yogurt before it was popularly marketed. Like the ice-cream man's treats, I even put "hidden" surprises inside. I know that this may not compare with being with "the gang", but it was done with love and the knowledge of our limitations.

Looking back, I realize that many of our habits were typical of a family unit of the fifties, while living in the midst of a society, rapidly changing to the value systems of the nineties. The "me" generation was already upon us. Two parent families with mothers at home were becoming increasingly fewer as the economy and society dictated that two salaries were necessary. Working mothers were becoming more the norm than not. We chose, however, to remain the type of family that we were.

Using good money management skills, I was able to realize a great deal of savings; on auto use, gasoline, clothing, social affairs, meal planning, creative cooking, and home repairs of all types. To me, this challenge compared favorably with receiving two incomes, a situation where there is neither time nor energy to do those tasks and neither to transmit the values that I wished to pass on.

We had much to share...baked goods from scratch, hand-crafted items, family visits, the joy of our accomplishments, and...love. I was able to paint bedrooms, make simple clothing, and know the satisfaction of repairing the intricate mechanisms of complicated toys. No way did I feel less like a woman or that a part of me was being denied or overlooked because I was an "at-home mom". Others may have thought that, but I did not.

My plan had been to give the children a solid foundation in the skills necessary for independent living while gradually preparing myself for re-entry into a professional line of work. I knew that when my husband reached retirement age, the children would be in college. I wasn't sure just how that was going to work out financially, but I knew that my long-range goal would be to return to my professional career in some way and that I had a good solid work basis on which to build. I began to do just that. When the opportunity arose for me to be a tutor in the Project Voice "Right to Read" program, I volunteered, tutoring first-grade students with reading difficulties and happily did so for three years.

The children were growing and there were a lot of fun times. But there were times when I yelled at the kids more than I wanted to - when I felt "housebound" due to my husband's long hours or when I was ill and there was no one around to care for the children. I viewed those times, sometimes with remorse, but also accepted them as the reality of life that it is. On the whole, I felt that I was a good mother and proceeded to continue with my individual development, as well as oversee that of the children.

A Time for Mom
The Beginning of "Holism"

In time, the children's patterns and schedules showed some basic trends. I was able to enjoy myself more and to develop my own personal and professional goals. I had been appointed to and was now serving on the Board of Trustees of the local library system.

Increasingly, I was using opportunities as they presented themselves, to re-establish and advance my professional skills and status. I was still helping out at the Boys' Club in various ways. Youth problems were beginning to become a focus of society with the development of federally funded programs for them. I had already volunteered in the "Right to Read" program in the school system. Now I decided to tackle a more difficult endeavor which, if successful, would add to the value and diversity of the Club's existing programs. I became familiar with the Reading is Fundamental Program (R.I.F.) a federally funded program, which I had heard was being conducted in the school systems. Its purpose was to promote motivation for reading, while developing the various skills for reading. I obtained applications, wrote the proposal, received matching funds and carried out the program. As part of it, I organized a local community group, consisting of parents and professional consultants in the field. I designed the program so that parents would run one session each and the consultants would act in the capacity of advisors. The program was successful. Not only was I enjoying these endeavors, but I was reactivating and updating my professional skills and repertoire.

All the while, I continued my research of how various nutrients affect the functioning of our bodies, and I continued to seek answers and solutions for Karen's problems, which, although controlled, still existed. Furthermore, I began to think of incorporating this data somehow into my professional nursing career.

Meanwhile, my life practices were casually and spontaneously developing into a pattern of practices and habits which seemed to be keeping me healthier. With some of the mothers in the apartment complex, I attended adult education programs at the local high school and learned the full routine of the U.S. Air Force exercises, (which I still do each morning upon arising). I began taking various vitamins and mineral supplements. My research about Karen had taught me a great deal about the deficiencies of my own system. As I helped her, my own health was improving dramatically. My psoriasis was gone. My energy level was higher. I didn't get sick as often and when I did, I recuperated faster. I cautiously decreased and later discontinued my thyroid medication. In time, I was found to have a "normal" thyroid reading, for the first time since medication was prescribed about 18 years previously-- even though I was told that I would have to take it my entire life. I did not discuss much of this with my physician. I was not sure that he would accept or approve all of these practices as having the benefits that I knew I was receiving from them.

I again felt the "free spirit" that I had often experienced in my earlier, less hectic years. I was bike-riding and swimming frequently, activities that I began when the children were small. I rode my bike in the early morning hours on the parkway and enjoyed the solitude that it offered....just me and the seagulls. I loved the sight of the bay on those quiet mornings. At that time of the morning there was little traffic to distract me from the sight of the white swans swimming against the contrast of the blue water and blue skies with the green grass beyond. It was enough to give anyone a morning "high" that would last all day. I did not know that I was daily practicing what would later be known as forms of "visual imagery". I did not yet know that this helped the secretion of pleasure-producing endorphins. Nor did I know that swimming and bicycle riding were going to be among the now popular aerobic exercises which would be encouraged for health reasons. I did not know that I was carrying out what would soon be known as "alternative" methods of health care, important to health promotion and the prevention of disease.

I could not afford the cost of "Transcendental Meditation" classes that were then being talked about and rapidly gaining in popularity, but I did take time out once or twice each day to practice my own form of "meditation". I set aside one or two twenty minute periods out of the day to let my mind wander until it seemed that all of the "clutter" of the day seemed to be organized, sorted out and "put away". I took time out periodically to strum the chords on my old guitar, an activity that brought me comfort and pleasant nostalgia. I took time to smell the roses and began to plant flowers. I did not know that I was "practicing" all of the wonderful therapies of nature or nature's musical meditation. I just lived and felt "balanced".

I was active, re-organizing and up-dating my career skills, enjoying nature and engaging in inexpensive, healthful, pleasant diversions. I did not know that, along the way, I had put together for myself almost all of the elements that now comprise a holistic health regime. I only knew I loved the beauty of God's world. The union that I found with nature could be captured in the moment and I knew that I belonged there... with all of nature.

A Program that Works

Outdoors was enjoyable for my children all year round. They participated in all sorts of games, rode bikes, roller-skated, and played in the yard games with their cousins and neighborhood children. In winter, they loved making snow forts, snowballs and Karen particularly loved making "snow angels". The lake where Grammy and Grampy resided provided a place for them to learn to ice skate and to sled and mini-ski on the wooded hills. We had good old-fashioned times with families and friends.

But, we still lived with the constant reminder that we had to function in accordance with Karen's heath limitations. Experience had taught me that not to do so would be unwise. I had long-since learned which types of situations, foods and environments would result in her becoming run-down, still leading to prolonged bouts of bronchitis, flu-like symptoms, or other symptoms of illness.

But we had made much progress. With time, I had gained an increased understanding of her problems and through experimentation, devised an activity schedule and diet more varied and compatible with her social needs. I found, that I could supply some of the valuable nutrients from foods that had been originally omitted from her diet by changing the way they were cooked or prepared. (She could tolerate eggs now if they were hard-boiled and eaten no more than about twice a week). I was able to incorporate many of "Auntie Muffie's" recipes, made from the traditional Italian food staples - peas, beans, pastas, corn meals and chicken seemingly without problems. These blended nicely with a diet mostly of home-made breads and muffins, rice dishes, broiled meats, sandwiches consisting of slices of cooked roasts, dark breads, yogurt, simply prepared pasta and plain fruit desserts; in essence, all foods made from basic, natural ingredients and without preservatives.

Similarly, I had found a brand of milk that had pre-digested enzymes. This finding was most welcomed, as milk drinks she had previously were ones I had made from diluted coffee lighteners. That was also the milk substitute that I used in cooking. Now we were able to deal with the problems of the "lactose intolerance". There was a valid, scientific name for that now and an explanation for the regurgitation in earlier years following the ingestion of milk. I could now erase the term of "nervous mother" for that problem.

The problems of high cholesterol levels, particularly in children, were not the wide spread controversial topics of public interest that they are today. Had they been so, perhaps I would have realized that, her diet seemed very much related to those to control high cholesterol levels! But, we had come a long way in keeping Karen's symptoms to a minimum and as we did, she became more free to live and play in a balanced way.

The children continued to gain rewards for their accomplishments. I was happy for them, but never took the compliments of their glories personally. I tried to share them with all of those who had provided a part in the realization of those accomplishments. I tried to keep the balance of the positive and negative factors that were influencing our lives, and each day I took a few special moments to put things into a healthy perspective.

I maintained my passion for life, which I always tried to share. I found my "soulmates" in music and dance, and in the souls of the many others whom I had met along their journey in life. My philosophy reflected a sense of the constant flow of life... that what is received is passed on, keeping only so much as one needs; and to make positive whatever negativity comes your way. I had a sense of compassion and commitment toward my fellow man. In solitude, there was a depth in me that no one seemed able to touch. I had walked a difficult journey in life, and was certain that I could guide my children in theirs.

Talent Develops

Karen was doing very well in the gifted program and loved it. Being bussed to another school district daily presented no problems. She enjoyed the new school and made friends easily. Although they were from another area of the city, they were able to meet one another after school and on weekends. She took part in after-school projects and continued dance lessons at the studio where she had attended since three years of age. As her talents continued to develop, I realized just how much she loved dance. (I knew that it held some special meaning for her since I had attended the routine "mother's visiting day" when she was six years old, at that time, I had noted the discipline that the ballet routines required and that she also had to learn the French terms. I remember thinking as I watched, "this is really serious stuff for a little girl". This serious dancing would never have been for me at her age, but it was "her love".) Her talents and her high level of motivation in the development of them had progressed in the new school. I watched her seriousness with admiration and encouraged her to always pursue her dreams.

My son was now in third grade. His talents were continuously obvious. Every favorite cartoon character, tiny birds, pictures of historic events; the Mayflower ship and Indians - all well-defined, and detailed drawings, charcoals, and paintings began to appear on bits of paper, large papers and scraps of cardboard, everywhere. He had developed his own innate abilities. And there was definitely an artistic quality about them...so much so that we enrolled him in a local art studio, in addition to the accordion lessons. Once again, he helped with expenses by "chipping in" from his allowance. The owner and manager of the art studio were very impressed with his talents---and he flourished in this area, as well.

Meanwhile, although I kept my distance as a parent within the school system, I was informed me that he, too, would be referred to Project Gifted.

My husband and I did not have to deliberate long to decide that he, too, should have the same benefits that Karen was experiencing. She was enjoying the school in which the gifted program was located. We were satisfied that the teaching staff worked very well in integrating the project needs with their regular classroom curriculum. They were capable, loving and seemed to provide just the right amount of challenge and appropriate amount of help for Karen. Ray, too, was screened and chosen for the gifted program.

Things would be somewhat different for him, though. He would be able to ride the school bus with his sister and, whereas, Karen had gone alone, there was more than one young man chosen from his school district who would be going with him each day. That meant that later on, he would attend junior high with at least a few peers, with whom he was already friendly. It seemed that he would have nothing to lose and much to gain. He, too would go to "Project Gifted"

I began to see an increasingly individualistic young man growing up. Many mornings as he went off, he wore the then-popular, multicolored bandana tied around his upper leg. It was obvious that his creativity and sense of humor were very much alive. I often wondered how he was ever fitting in with the more serious-minded, students, but this was his style. The bandana, in itself, was harmless and I had learned by now of the "individuality" which seems to characterize many of the "gifted" and "creative" people that I knew. And so the bandana ran its course.

Karen enjoyed him and the two children remained very close for a long time, even though they demonstrated the usual sibling rivalry. They made good times together in their shared love of humor and creativity. One summer they directed and produced a performance which they put on for the neighborhood children. They made puppets, sewed the clothes for them. made stage-props and planned a presentation of the children's story, When Ira Sleeps Over. They made other puppets which they used for a comedy portion of the program and elicited audience participation. At intermission, my son played his accordion and they sold home-made lemonade. It was not as common now for children to be producing and carrying out these kinds of activities for their peers as it had been in years past, but all of the children enjoyed it...mine in presenting, and the others in attending. It provided them with an early experience in entrepreneurship and fun.

Karen continued to develop her talents. She loved the hours spent in Project Gifted. The arrangement there was that all students attend regular classes at their own grade level, with a designated number of hours spent each week in the "gifted" classroom. There a teacher, specific for that program, conducted innovative "enrichment" programs. (It is to be noted that these were indeed "enrichment" programs, not the academically advanced programs which were at that time taking place in the higher grades and being called "accelerated".) There were classes in mythology and logic; current topics and philosophical issues were also discussed. Later, the inclusion of various projects (video, performing arts) fostered the development of creativity and the increasing use of technology, the making of videos, and the inclusion of a variety of the performing arts.

(There was one particular young man in the class who matched my daughter's love and exuberance for the creative activities. They spent long hours developing many projects together. They formed a close and lasting friendship in which they shared many things. In the near future though, each, in their own way would learn about and suffer, the effects of the unique sensitivities that often accompany many of the children who are known among the more "highly gifted".)

Karen was now in sixth grade and had functioned fairly well health-wise. By maintaining the program of diet, rest, desensitization injections, and other limitations, her "allergies" were kept under control. There was only one recent indicator that the original problems still existed. A bout with a winter flu turned into a three week absence from school. That was not an unusual occurrence, in view of her past, but she was older now. Missing school on account of illness was much more notable. So, at the end of the three weeks, I sent her back to school, still exhausted and looking so. This time it was the teacher who sent her home again for further recovery. She saw and commented that Karen's return to school was premature. All I could provide was the usual explanation that this was a "normal" course for Karen following an illness and so there were no particular consequences at that time. Her pediatrician and allergist, who were still working together, seemed to think nothing of this pattern, knowing it as usual for her. Following this long recuperation, she returned to school and there were no penalties or unusual comments. The teachers had come to know Karen and me, and to them, the long period of recovery was a matter of fact. These were honest, trusting, caring personnel, working with students and parents for their welfare. But, deep inside, I knew that the simple acceptance of long absences from school because of her condition could not last forever.

After her recuperation, Karen's motivation was just as high as ever. She seemed to be enjoying all aspects of her life, but, most of all, she continued to enjoy dancing. She heard that the studio where she took lessons was going to offer for the first time, a more difficult, classical ballet class. She pondered the thought of taking it and I marveled at her maturity as I listened to her decision-making process. She concluded that she would wait one more year, stating that she was "young" and wanted to be sure she was "ready" for it.

She seemed to be serious about all of the performing arts. She wrote a song and piano accompaniment for a class project.

The name of the piece was titled, "Changes for the Future". It was a very serious theme, addressing the issue of peace for our future and ideas on how to attain that peace. This was in fifth grade! But, the seriousness that I so often saw in her seemed to be balanced with humor and the good times that we all enjoyed.

Little League season came and both children enrolled. This was to be her brother's primary active pastime, balancing the pursuits of his artistic talents. He was a good third baseman and dearly loved being a part of the team. Yet Karen was not to be outdone. With my reluctance at her taking on yet another activity, she joined. Did I need to question that she had talent in this endeavor, too? And even with all of these things going on, her intellectual pursuits did not suffer. At the end of sixth grade she went on to enter--and win--the city wide Spelling Bee. We maintained her health regime and she had flourished!

But, while she was busy developing her talents, she was nearing puberty and developing a keen sense of self-awareness. She often felt and expressed that she was "different" from many of the others. (In my innocence as a parent, I knew nothing about this thing called "gifted". I did not know that a sense of "differentness" and isolation is so often characteristic of gifted adolescents. I just listened.) She did not seem to share in much that was common to other pre-adolescents in sixth grade, about to enter junior high. She was a perfect example of the fact that we all develop differently and at different times. While their interests had begun to be consumed with the things of adolescence, her passions remained those for schoolwork, the creative arts, home and family life. That was no problem for me. But none of us, in our wildest dreams, could begin to fathom that, later on, her world would become so troubled, causing her to shun the peers whom she loved, and to sob uncontrollably "It's not nice to be smart. It's not nice to be pretty. It's not nice to be 'good'."

What subtle messages are implicit in society that one so bright and gifted should have to experience the pain and conflict of whether to develop or hide precious traits and abilities? Was her song "Changes for The Future" to be a prophetic task that she and I would have to accomplish?

On to Junior High

Karen looked forward to beginning junior high and to what she thought would be the rekindling of friendships with the peers she had left in third grade. But it was different now. Naturally, the classmates that she had known in her former school district, had changed during those four years. Their friendships and values were changing, as they themselves had changed. (They were beginning to explore and formulate different sets of values.) As sociable as she was, Karen found that it was not easy to resume the friendships. Students coming to the junior high were from many schools but did not include students from the district where she had attended Project Gifted. So, at a time when a peer group was so important, she was like a stranger in a new land, even though the new structure was within walking distance.

Our decision to send her to the large public school, rather than a private one, was based on the thought that it was advantageous to learn to work and play with a wide variety of people. The other factor in our decision was that we thought that college might well be a possibility for our children and that any funds that we could accumulate for education would be more appropriately designated toward college expenses for our two talented children.

Karen brought with her the same enthusiasm for new adventures and for learning that she had always exhibited. She joined the band and made attempts at establishing new relationships. Her presence, character and exuberance were quickly noted by the head guidance counselor. Even though he had hundreds of children under his supervision, he seemed to be "tuned in" to each of them. In his long years of student involvement, he had developed a sensitive heart and eye for existing or potential problems. Later, he was to play a most important part in our journey and demonstrated clearly that he was every bit indeed, what his role implied - a guidance counselor. Along the way, we would find others who would have titles that implied that authority and knowledge, but whose knowledge was indeed limited in scope.

I watched Karen's continued enthusiasm and eagerness to experience life in this new endeavor. Few of her peers seemed to put the same emphasis on the same things that she did. Boyfriends, girlfriends, and other adolescent interests were higher on almost everyone else's priority list. I could understand her expressed disappointment and the conflicts of that fact. These were the teen years and peer interests were important. In addition, the burden of having to follow a strict dietary regime, and obtain necessary sleep did not correlate with the desire for unlimited activities and lack of sleep characteristic of those teen years. These factors, combined with the approaching hormonal changes of puberty would soon take toll. But, for now, life was going O.K.

I continued to pursue my personal and professional goals, and had found part-time work as a mentor in a federally-funded program. Its aim was to teach basic reading, writing and math skills to adults. The students had a wide variety of skill levels and backgrounds and exhibited much diversity in physical, social, emotional and cultural characteristics. The program provided them with the opportunity of "moving up" from the lowest grade level of performance in which they entered, to the highest level that they could attain. (Thus the program's title, "Moving-Up". For some, the level of achievement was that of college enrollment. Others went on to appropriate job settings. I thoroughly enjoyed my assignment of working with the lowest level adult readers. I taught them very basic skills of writing a check, reading a menu or whatever their needs were. It was truly satisfying and challenging.

This re-entry into the work world confirmed for me that my area of strength and satisfaction continued to be that of rehabilitation and of working with people who could develop their potential, given encouragement and the opportunity to do so. I enrolled in a Master's Degree Program in Rehabilitation Counseling, the completion of which would bring me a Master of Arts Degree in Agency Counseling.

A Camping Experience - A Turning Point

Even with the addition of new activities and responsibilities, we managed for the most part, to maintain our family value system. For a long time now, Karen had not experienced any serious health problems. With my urging and constant guidance, she was maintaining her program well. I felt very strongly that Karen's enjoyment of relative good health was due to the fact of the regime that we established some time ago and was now being adhered to without question. She had been taking large amounts of vitamin supplements, minerals and amino-acid preparations daily, and carried out the other precautions which lessened untoward reactions to things in the environment to which others did not react. These seemed to have had a big impact on her "healing", and kept her able to function well. She took the supplements with her to school, on vacation, everywhere. She ate yogurt on a daily basis. She took a brown-bag lunch of natural, low-fat foods to school, instead of eating school lunches. She had a schedule of rest and bedtimes that afforded an appropriate amount of sleep. As long as she stuck to prescribed bedtimes with the scheduling in of some later-to-bed nights, for special occasions, she could maintain a fairly "normal" schedule for a very bright and motivated young woman.

But she was now in junior high. It was difficult to maintain the restrictions of such strict dietary and sleep regulations. My husband thought that I was putting too much emphasis on having her follow so strict a routine, since she was doing so well. To my way of thinking, she was doing so well because of the routine. We were in constant "hassle" about this. I gave her as much freedom as I could, in as many areas as I saw fit, but I also saw indications that the original problems were still there, and controlled by the established health maintenance program.

I always knew, however, that the time would arrive when I would not be there to oversee, to prompt, to enforce these regimes that had been a lifelong fact of her life and which had seemed to provide her with good health and energy. But those were "my" thoughts, "my" observations and the results of "my" research. This would soon be "her" life, "her" responsibility" and "her" body to regulate. She would one day have to internalize this fact and to realize the uniqueness of her own special problems. That day was coming soon. Was she ready? Could that happen now? Is there ever a "right" time?

The questions remained. Had her system changed? How much did she really need to adhere to those previous restrictions now that she was older? We decided that we needed to "let go" to find out.

Adolescence is a time to develop one's self-identity. It is important for teens to have the opportunity to develop in relationships outside of their familiar environments. I needed to let Karen grow in the knowledge of her own uniqueness, just as I had in all other phases of her life.

So, in spite of her numerous problems, (which were still being called "allergies"), we discussed the idea of her enrolling for a week in a summer overnight camp. She still loved adventure and newness with a passion, and, as far as we were concerned, she deserved a week in which she could take part in arts and crafts, boating, swimming, and all things that she loved to do. There was one small problem. No one she knew would be attending this camp or, for that matter, any other camp this summer. To go with someone else would have been a preference, but, true to form, she agreed to go alone. Anyway, we had an understanding that she could leave the camp at any time if she didn't like it.

Prior to her going, she and I discussed which parts of her

health maintenance regime she might be able do without, and what ones she might be able to continue on her own. We went to camp for a tour prior to the scheduled date. She was shown just about everything that would be of importance for her, including the nurse's station. We discussed her "allergies" with the representative staff and asked if she could be responsible for her own "treatment", basically the vitamin/mineral preparations, taken several times a day and a milk supplement, providing calcium. According to the policies, she could take only her anti-histamine, while there because it was an "approved" treatment. That could be kept at the nurse's station. She would not be able to take any part of the vitamin supplementation, nor even have them with her. They were not prescribed "treatments". The nurse would not supervise the taking of them - nor were they allowed to be kept unauthorized in the tent. That was policy!

O.K. We would chance that-- and the effects of her being without the other health benefits that she was used to-- (daily nasal irrigations, air conditioning, etc.). We also were willing to chance the possible effects of some other stress-producing factors, such as late hours, irregular meal times and environmental factors. She really had to be given a chance to live like the rest of teenagers.

Karen and I discussed the snacks that she could take, so that at least some part of her diet could be maintained. We talked about the importance of her using sunblock, knowing that unlimited exposure to the sun had also been a risk factor. And, for now, we would forget about the many and varied kinds of trees, molds and pollens at camp. We hoped that the weather would be somewhat dry and cool.

This was part of the growing-up process and I had to allow it. For her part, she was happy to be trying something new, eager to be outdoors and looked forward to preparations for a week at camp.

Her father, brother and I went to the campsite that Sunday

afternoon. Numerous other teenage girls were talking excitedly and seemed to be going in all directions at once. I wished for her that she could have been with a friend but such was not the case. She met the other young girls and found a place in the musty old tent, to set up her bedding and mosquito netting. Having done that, she prepared to settle down in what would be her bedroom for the next week.

Her cot was at the end of the tent next to the smelly, moldy tent flaps now down, because it had begun to rain. In fact, inclement weather had already been forecast for the whole week. We hoped that they were wrong. Shortly after, we left her there with the others. Tears streamed from her father's eyes and the rain poured down as we drove home quietly.

We thought about her that evening, as would most parents. I especially thought about the thunderstorms that we were having. It meant that the moldy-smelling sides of her tent would have to be down at least all night. I thought of the times she had stayed at Grammy's and about her long-term problems with humidity. But I also knew that she had to go through this "test" and hoped for the best. After all the day of "awareness" would have to arrive sometime.

That night we listened to the week-long forecast. Rain or showers were predicted in some amount for each day of that week. Although we received postcards, which sounded cheerful, telling of her adventures of "K.P." duty and the like, I wondered how she really was. Was she really surviving all these adverse conditions with no ill-effects? I would be thrilled if that were the case.

It seemed like a long week by the time that we ready to go pick her up that following weekend. When I first saw her she looked so different. She looked so tired. She spoke little, (which was not like her), and seemed just totally exhausted during the ride home. (I would not know until later, just how totally drained of energy that she really was and how the many stressors had taken their toll on her). Nor did I know that she would go on to develop symptoms of a condition which nobody would be able to diagnose because the symptoms had not yet been described medically.

The first thing that she did when she got home was to immediately take out mixing bowls and begin to cook. I thought that odd, but left her to her own activity. I observed her as she devoured what she had made. Then she sat out in the yard in a lounge chair, which she often enjoyed doing in the summer. But this time, she was so quiet and sat for so long...several hours. She was not sleeping or reading, just sitting gazing into space. Again unusual. Her mood seemed to be more of a depressed one than just quiet or exhausted. She did say, when asked, that she had enjoyed the camp experience.

Later on, however, we found that she did not like having gone on this particular adventure without a companion. Almost everyone there in her unit had gone with at least one other person. And there were other things. Even though the anti-histamine tablets were available for her, she still experienced severe stomach cramps. When she went to the nurse for assistance, other girls in her unit said that she was not really sick and was just going there to get out of chore duty. She explained how badly that had made her feel because that was never the case. We knew her to be highly conscientious, able to withstand a lot of pain and discomfort and an extremely willing team member. This must have been very difficult for her. Yet, this was to be the beginning of the long and difficult journey of an honest, sensitive, young woman learning about and facing the immaturity and insensitivities common to so many people in our society, when they have not walked in the shoes of another.

Where is My Daughter?

What I noticed most in Karen that summer, was the severe lethargy that I saw in her, markedly contrasting to her usual enthusiasm. It lingered--and was severe. I tried to take all of this in stride without placing undue emphasis on it. I was mindful of the questions and searches for identity that characterize the adolescent period and aware of the distance from parents that adolescents seek. But this did not seem to be any of those. Even with those considerations, this was not characteristic of my daughter. Was it possible that this was the result of an "onslaught" of exposure to chemicals and elements that her body had truly not been able to fight? Could multiple "allergies" cause this much of a long-lasting after-effect?

The real "eye-opener" for me was when we were preparing to go on our usual mini-vacation at the end of the summer. We had always gone somewhere for a short vacation during the latter part of August. My husband's work schedule afforded a break between the summer and winter programs. This time we had planned to go to Wood's Hole to see the Marine Museum and to take an excursion to Martha's Vineyard from there. We had made reservations for an efficiency unit with air-conditioning. This afforded us the ability to use our own supply of natural food staples, keep expenses at a minimum, and have fun, while minimizing undesirable effects that Karen might experience just before the school year. This year we had even brought a special enzyme-treated milk with us that was now on the market. I was glad that we had these extra accommodations which could keep the "stress" factors to a minimum, especially now.

But there seemed to be other things going on with Karen

and I was not sure what they were. She was not the daughter that I had known, adolescence or no adolescence. There was something else.

On one of the mornings of our trip we went out to a restaurant to eat breakfast, usually a special treat for Karen, when we did, and something that she had always enjoyed doing. But, she only went reluctantly, with coaxing. Once inside, she was solemn and seemingly fearful. After every few bites of food she'd go to the bathroom. After several instances of this, I followed her, ever mindful of her changed condition and aware of the recently publicized problem of anorexia among teens. She was not an over-weight child, nor had weight ever been an issue. Therefore, I did not really believe that this was related to any form of anorexia, but I had to investigate. When I asked her what was happening, she said that she "just wasn't feeling well".

During those few days, she became fearful often, and would break out in a sweat. She refused to go to restaurants for the remainder of the stay. This was the girl who loved to "cheat" and truly enjoyed eating out whenever we could do that. Something definitely was wrong! When my husband tried to make light conversation about the beginning of school, which would reopen in just a few weeks, she just looked fearful and made no comment. I kept thinking, "where is my daughter"? Where is that enthusiasm? What has happened?"

She had always loved school and had already put in a whole year at the junior high. What was happening? I tried to brush it off as something that would pass but I soon knew that it would not pass quickly. When she seemed panicked as we approached the ship to go the Vineyard, I knew that there was something seriously wrong. She stood almost frozen on the dock, refusing to move to get on the boat. After we made several attempts to persuade her to get on, she was able to make the trip with us. She was able to verbalize the fear that something would happen to us on the ship. This was *very* unlike Karen and her sense of adventure. It was obvious to me that she was experiencing something, which I did not understand. She was beginning to have "fears" instead of the "treats" that we thought she was going to have. I couldn't believe that this was happening.

I began to wonder about the beginning of school. Could she "recuperate" by then? Would she "come 'round' by then? I truly believed that her failure to adhere to the routines that had brought her to health and had maintained it, was somehow very much related to these behaviors. They had noticeably begun with the interruption of her health regime and the additional stress factors that she experienced at camp. I could only watch and wait. But, I somehow knew that this was, in some way, related to all of those symptoms that I had noted in Karen since birth. The "condition" or "conditions" had not been thoroughly understood or adequately explained, I thought. In my heart, I believed that we had kept the symptoms under control through diet, rest, immunizations and environmental precautions. I tried to convince Karen and my husband that she should resume at least some part of the treatment that had worked, but I could not convince them of my convictions.

Her father wanted ever so much for his daughter to be "normal". I wanted her to be happy and symptom-free. We did not share the same knowledge about some of the ways that allergies can manifest themselves. Without this knowledge and the belief that her former treatments were of the utmost significance, my husband was like an outsider looking in, not understanding what these "monstrosities" were that were invading our family life.

I so much wanted this adolescent to have freedom of choices in her life, but the freedom of choices had to include self-knowledge and self-responsibility. Only then would she be able to truly make "informed" choices. I could not force my beliefs on her, even though they were based on research, experience and careful observations of my daughter. Nor did I want to. Mothers can urge, explain, love and teach, but not "force". I was over-ruled!

What to do About School

Two weeks later, school began. I spoke to no one about my concern. But obviously the thought of her not attending school was ever-present.

On the first day of class, I sent both children off as I would have any other school year...with my blessings, love and support. I could see the effort that it was taking for Karen to go off, but I said nothing. After about two days, she returned home one morning shortly after having left the house. She did not need to say anything for me to know that something serious was going on. I had seen it in her face and actions for the last two days and weeks. But now there she sat, in the big stuffed chair in the living room. Attempts at persuasion were ineffective. She could not verbalize what was happening, but it was obvious to me that something "really big" was going on that was keeping her from going to that which she had always loved.

A day of rest and conversation to attempt to "find" the problem was unsuccessful. There were many tears, but she still was not able to explain what was happening. I knew that this was not a phase or something that would just fade away. Something had caused this obstacle. It would be necessary to define it and deal with it appropriately.

Although she tried many mornings to go to school and remain there, she always returned home shortly after she had left. She was not able to accomplish whatever it took to make the four blocks to school. On one of those mornings, after she returned home, she sat in the big chair in the living room, almost despondently and said solemnly, "Mom, I'm not the kind of girl that would run away, but I just can't go to school. I hope you're not going to be the one that has to make me go because if you are, I'll hate you." I thank God for the wisdom and courage instilled in me that enabled me to say, "No you won't, Karen, you'll love me".

During our talks, I found that she became fearful and anxious and had experienced outbreaks of rapid heartbeat and cold sweats. (That gave us further confirmation that she really did need those mid-morning and mid-afternoon snacks which had always been a part of her program.) I deeply wished that she had resumed the allergy treatment regime. But if self-discovery about this were to be part of the problem and the solution, then I had to let it happen.

I called the head guidance counselor and told him what was happening. He recalled who my daughter was and remembered her as an exceptionally vivacious student. He listened to my report of what was transpiring and agreed that we would work together on what he thought would probably be a small temporary problem.

The plan was that I would try to send her off each morning and that he would watch for her arrival. We would be in contact each morning to discuss the outcome and plan for the next day. He would interact with the teachers as needed. When, after several days, no progress was being made, and she was not successful in getting to school, he, too, knew that something more needed to be done.

He suggested that, perhaps, it was just some small "adolescent" problem that she did not want to discuss with me. I respected this opinion and agreed to visit a local counseling agency to rule out or confirm this possibility. The guidance counselor knew the staff there and had often made referrals for others. He was able to get an appointment for someone to see her immediately. Frightened though she was, Karen agreed to have a private conversation with a counselor. My husband and I waited in an outer room.

The counselor spoke alone with her for a very long time, but apparently was unable to gain any insight from the interview. He then called me in and asked me about her social habits. I told him of how she had been highly sociable from a very early age. I told him of her ability to initiate new friendships and how she had always maintained an open mind for new experiences and of her willingness to attend Project Gifted. In fact, she had been one of the few children in our plat who, during elementary school, had always played with large numbers and varieties of children. In his search for information, he asked if she had been there lately. No, she did not hang out with that gang, although she knew some of them from school. There was a different group living there now. Where there were once small children, who were well-supervised, there was now an older group and a more transient population. The large lot had become a hang-out for teens who were out late at night with their boom boxes blasting. I did not wish to encourage that type of recreational activity. He seemed to have no indication of the problem and we left there with no apparent resolution and no word of another appointment.

Later that afternoon, he telephoned us to say that he had been thinking about the discussion earlier in the day and believed the problem to be one of "trust". He was nearly "convinced" that there was some prejudice in our family because I wasn't letting her "hang" with the large group of teenagers in the apartment complex. Interesting, I thought. Had he found out something that I didn't know? Was something going on that I did not know about and that Karen was reluctant to tell me. Had she really wanted to "hang" with these kids? And even if she did, shouldn't parents have a say in where and with whom they would like their teens to "hang", without such consequences? This was a matter of parenting, but, to say that there was prejudice, was like saying that I was a zebra! It was just so idiotic!

I did not have answers. I was looking for help. Did this "counselor" possibly discover or uncover issues from my daughter about which we did not know? Did he really think it beneficial during these teen years to let her mingle with the group in the apartment complex? What if it was some sort of teenage rebellion that she was experiencing? I wasn't so sure that had anything to do with what was going on but if letting her go to the apartment complex to "hang" some evenings might give some clues as to how we might help, I would try that. After all time was going on into the school year and she was *not* going to school.

Going to the "apartments", (as we called it), would, at least serve the purpose of her getting out of the house and afford her with some mingling of some of her age group, (socialization which she was not otherwise getting right now.) She was even beginning not to even want to talk on the telephone with school friends because she was not able to tell them what was "wrong". So from a general socialization perspective this venture might serve some purpose. I did not relish the idea of her hanging out late at night with this mixed group of kids. My heart broke as I watched her go out each night after dark to the group next door. I watched her getting more and more tired. This would not make her go to school. Was this what parenting was about these days? I thought not. After several nights of this, I felt satisfied that I had given even this opinion a fair try.

There was another thought that this same counselor had expressed in his follow-up conversation with me. There might be a matter of "trust" involved but I couldn't "buy" that. Nor did I honor his request to stop by his office to sign a form. When I inquired about the nature of that form, he said I needed to sign on the form that I had experienced problems with substance abuse. Only by my admitting to some sort of substance abuse problem could the office visit be legitimized for payment. I was flabbergasted! No way would I agree to something like that! If some type of counseling were needed, then I would find it elsewhere and pay accordingly. I had, at least, followed the suggestion of the school guidance counselor in seeking and obtaining an "outside" opinion. But, this episode was to be the first of many guessing games, attempts at "cures", and reasons to blame for what were basic physiologic problems accompanied and aggravated by a gifted young woman's struggle through adolescence.

As I listened to each of the "helpers" that we sought or who were assigned to us on our journey, I learned that each one's insights seemed to be relevant to their own situation in life and not necessarily to the issues at hand. Moreover, whereas my concept of counseling had always been that "helper" and "helpee" work together in an effort to assess the source of problems and try to find solutions for them, I found that most of the would-be "helpers" seemed to think that their roles and the knowledge that they possessed were those of absolute authority. They were the "experts", or so they thought.

In retrospect, I realize that I tried to give the professionals involved the "benefit of the doubt", as they say. I did not have the answers; maybe they did. I tried to give them the trust and respect for their competence that I like to have. As a professional and parent who respected my growing daughter's privacy, I adhered to what I knew, as a code of professional ethics. I never asked my daughter or her "counselors" what they were discussing. If there was, indeed, a problem that involved me, in some way, I wanted my daughter to have the distance from me that she needed to resolve it, and for me to be involved in that process, only as needed. I was not looking for "blame". I knew my weaknesses as a person and parent, I was looking for help with something that I did not understand. In retrospect, I think that we all would have benefitted more if I had not brought my concepts of integrity and professionalism into these interactions, for I was to find that my idea of "professionalism" did not match that of the others. And I should not have let them be the "expert" in an area where I was just that. But did I really have that choice? She was still under the dictates of the systems.

Another Look at Physical Aspects

In one sense, there really did not seem to be a need to investigate anything more in the physical realm. I had been informed during the last several years of what Karen's problems were and what we were supposed to do for them. We had been working with most of those symptoms since her birth. Simply put, she had "allergies" and sensitivities to environmental conditions. The symptoms now seemed to result in complete lack of energy and motivation for usual activities. How much was physical and how much was social or related to adolescence? We continued to explore both.

In spite of her having had exams by both the allergist and pediatrician just prior to camp (two months earlier), I decided to investigate further her complaint of "stomach problems". That had been one of the most problematic just after camp and had exacerbated during our vacation. She still had not resumed eating properly, although previously she always had a healthy appetite. I believed her when she said that her stomach was "just too upset to eat". We went back to the pediatrician. Karen, he and I had a long conversation. In a fatherly way, he tried to uncover what might be bothering her. He did not seem unduly excited and attributed the problems to a "nervous stomach". He wrote a prescription for medication that we could use if we chose to. I listened, but Karen and I both agreed that, although her stomach was bothering her, it was not a "nervous stomach". We held on to the medication, but did not use it.

As follow up in response to the complaint of extreme fatigue, tests were done for mononucleosis and low blood-sugar, but these came back negative. A re-examination by the allergist confirmed that the stomach-aches and other "stomach" conditions were, most likely, due to the "allergies" and left it at that...more confirmations, but information which seemed to mean nothing to anyone, but me.

In the meantime, I had begun to think more about this word "gifted". More than once since fifth grade, Karen had said that she had felt "different" from her peers. I remembered having read a book at that time that was recommended by one of the teachers in the gifted program. It was titled, *On Being Gifted.* It described some of the isolation and difficulties that *some* of the "gifted" experience. As my initial introduction to the subject, the book gave me insight into the fact that some who are called "gifted" experience real difficulties due to their many sensitivities. Could some characteristics associated with "gifted" somehow be part of Karen's problems? And who would know, anyway?

Whatever the word meant, it was beginning to become an important subject. It was now a topic for T.V. talk-shows and now there were four schools in the state which had developed programs for the gifted. Although I kept this in mind, it did not seem to be an immediate consideration, probably because I did not know what it really meant. At the present moment I was more concerned with my daughter's health rather than with a topic called "Giftedness".

The days went on and she was still out of school. I kept in constant touch with the guidance counselor who, himself, was now quite concerned. He contacted the Juvenile Division of the Department for Children and the Families (DCYF), who were often called upon to deal with school absences. He thought that there might be someone there who could help us. After all, we were now nearing a month of absences and all we really had for a diagnosis was a "nervous stomach". I saw that she still seemed to have periodic sweats, and what seemed to be episodes of severe anxiety, but how would that make anyone stay out of school for so long? No one was associating any of this with vague symptoms of low blood sugar which were probably exacerbated by the now decreased appetite. No one, (except me, apparently), would associate or even suspect that any of these symptoms could have been associated with faulty systemic problems that had been controlled through the use of high vitamin-mineral, amino-acid supplements and of which she was now going without. Maybe it was "psychological". We would talk to the worker from DCYF.

Within a day, a pretty young woman in her twenties, appeared at our door. She was a personable, bright and likeable woman to whom my daughter was quick to relate. Ironically, her name was Karen! She was eager to try to identify the problem. Even though she spoke with my daughter until midnight, she did not learn anything other than what my daughter was telling us, her parents. The worker told us that she would be in contact with us throughout the week. She came to the house and talked with my daughter daily, even volunteering to stay with her at school all day if that would help. Her attempts too, were unsuccessful.

Another Facility...Another "Treatment"

We had exhausted the usual routes trying to get verification that Karen's physical problems could cause such prolonged symptoms. Although I was sure that the causes were not the ones which we usually think of as "psychological", I would consider going somewhere that offered more comprehensive services. There was a well-known pediatric facility very near to our home, highly reputable for its work with children and their families. I suggested to the young DCYF worker that perhaps we should go there, since so far, we were making no progress. Karen was still not returning to school. For sure the professional facility noted for their work with children would be able to sort this all out, I thought! If marital differences were part of the problem, then we would try to resolve these. If chemical sensitivities or the attributes of being "gifted" were distressing her, then these issues needed to be resolved. At any rate, it seemed, that we needed the benefit of professionals who had the combined knowledge, skills and abilities to dissect and re-integrate the physical, mental and intellectual components of this puzzle! And we needed someone to start the process of home tutoring. Time was rapidly passing and she remained out of school despite our many attempts to have her return.

As distasteful as these kinds of interaction were for my husband, he did agree to go to the facility to see if we could find a solution to the problem. Following intake, we talked to an assigned counselor for a long period of time. Eager to help my daughter, I reported *anything* that I could think of that might be of significance. I was very honest. I reporting everything that I could think of that might have caused the problem. (Little did I know that later this honest revelation would be exploited upon.)

But there were other things that I thought that they should know, as well. They needed to know about how intensively sensitive Karen was and about her vivid imagination. Last May, she had come home from school very distraught. They had shown a movie in school that afternoon, portraying an accident in which adults were instantly killed as a result of someone's drunk driving. She was highly affected by this. She worried, intensely and aloud, "What would ever happen to me if something like that should ever happen to you two?" At the time, I had listened to her fears and comforted her. I did not find that worrying to be unusual or disturbing. I knew what a sensitive little person she was. I only wished that she did not feel things so deeply. Thoughts of death, particularly, of parental loss, are often a concern for adolescents. And it was understandable that such a sudden loss would be devastating for Karen. Not only did she love both of us dearly, but also she knew that we were helping her through her many problems and sensitivities. Knowing this, I tried to honestly explain to her just what would happen. Her father and I had made a will and arrangements for the guardianship of she and her brother long ago. We hoped that such a tragedy would not befall us until many years hence, but, nevertheless we had made provisions for them. I also expressed my strong belief that the chances of that happening to us in any near future were highly unlikely. She seemed satisfied and reassured.

I reported this incident to the therapist so that she might gain some insight into Karen's sensitivities and how the misfortunes of life so deeply affected her. But the therapist put this into a very different perspective. In her theoretical rhetoric, and interpretation of the incident, she said that the real issue was that Karen was repressing thoughts of wanting "something bad" to happen to us because we were so strict! I could not believe this and thought, "why had I been so honest"? It was I who had reported the aspects of our life which were not so pleasant. Certainly, there were occasions when my husband and I lacked patience. There were even outbursts at times. But, want us to die...no way! I could not believe what I was hearing! I really felt bad for that therapist and her outlook, for at that point, she had not even interviewed my daughter!

After a few sessions, it was obvious to us that the therapist was not on the right track. One of the recommended treatments given along the way was for my husband to "tickle" my navel. And these were to be the "authorities", the "experts", to whom I was going for "help". They made their conclusions and, in turn, reported their "findings" to the school.

I was told that the first report that went to the school system indicated that I was an "over-protective" mother. More labels were going to be added to the early one of "nervous mother". My husband, at this point, decided that he did not need this kind of "treatment" and would no longer attend "therapy". I knew that my attending was probably the only way that the "system" would accept the fact that Karen had a valid reason for not going to school. I hoped that they could help me find it-- so, I remained "in therapy".

As the weeks passed, my daughter showed no signs of being able or ready to attend school. She and I both attended "therapy" sessions weekly at the well-known facility. We were each assigned to different therapists who were supposed to be collaborating their efforts toward resolution of the problem-- a student internist for Karen and an older, "seasoned" therapist for me (the one whom I had been seeing since the intake).

A number of weeks went by. Karen was still out of school. The young worker from DCYF was making several visits to our home each week. She and my daughter got along very nicely. They were able to communicate on just about everything, it seemed, without reservation. But no progress was being made toward school attendance. In the meantime, I had been successful in getting some homework assignments. My daughter, bright as she is, readily taught herself those things which she did not already know.

The frustration for me was mounting, though. If we wanted the continued cooperation of the school system, there was no choice but for us to continue the "therapy" and they didn't seem to be doing anything. Karen vehemently opposed every attempt to go there. On one afternoon, angry at having to go there, she ran off into the nearby woods. With a heavy heart, I searched for over an hour and found her just as darkness was falling. I was able to convince her later of the importance of the appointments, explaining to her that we needed to go to that facility because the "therapists" were working with the school staff. That was the only way that we could keep our connection open with the school. Her love for school and desire for continued connection with it prevailed, but more problems arose.

The therapist seeing me, wanted Karen to meet with a psychiatrist. Karen, in turn, fiercely expressed to me that she did not want to meet with him. She did not know him and did not want to see him, even for an evaluation. She repeatedly, and adamantly, stated that she, Karen (the DCYF worker) and I would work it out and that we didn't need them. But we *did* need them in order to continue to receive homework. Loving school and trusting me, she finally, but reluctantly, agreed to meet with him.

When we arrived, I told them what she had told me about not wanting to talk to the psychiatrist, at least not alone. She maintained that she did not know him and did not want to talk to him. (Realistic thinking, I'd say.) Maybe I could be there with her while she spoke with him, she asked. That was her suggestion, not mine. But in no way was I prepared for the insensitivity which followed.

True to her fears, she was told she *had* to talk to the psychiatrist. Any attempts that I made, even to try to explain why she had said she was fearful and reluctant to talk alone, only added to my "over-protective" label. I was not to "interfere". At least I felt that my conscience was clear. I had acted on behalf of my daughter, relating to them the fears that she had expressed to me about not wanting to talk to the psychiatrist alone, yet knowing we had to go through with this if we ever wanted to get home tutoring. Once again, I was looked at and spoken to disdainfully. I really began to wonder if we were victims of some sort of a police state. Didn't they *really* want to hear what was happening or were they just following some pre-established set of rules and mandates?

Their interviews revealed that my daughter was smart, very smart and yes, probably "gifted". Because of this, an absence from school for her would bring no noticeable consequences as far as her grades were concerned or in her ability to catch up. Therefore, they could not recommend home tutoring. Her strong desire to learn would be the very thing that would get her to school, they thought. That was their "bait" to get her to school. They had missed the whole point again!

Here was a girl who loved school and was trying with everything in her to attend, and "something" just would not let her do that. We had come to them for help in trying to resolve that issue and they were failing us. Didn't they realize that it was the "immobilization" problem that we wanted help with? Didn't they "hear" anything?

The same pattern of disruption was happening in regard to her attending ballet lessons. She'd get all dressed, always wearing that same fearful, depressed look on her face, and we'd get in the car, to ride to the studio. Sometimes she was able to get out of the car, sometimes, not but never able to get into the studio. As I watched her crying and struggling with her feelings, it was most difficult to be an objective parent. I don't know how I was able to be so strong and continually supportive in my efforts at trying to break the spell, as day after day, week after week went by. It was always the same routine. I'd go through the motions of bringing her to the school.....and back; to ballet.....and back... ever with the hope that one of the times she would be able to go in and I could leave her there. But the "gripping" fear had taken its toll.

She no longer seemed to be able to go anywhere with anyone. She clung to me as though her life depended on the security that I could provide. This tore me apart. Trips that we planned with her cousin, whom she had always liked, would end up in much the same way. She would get all packed for the beach and, if lucky, get as far as the car, but never beyond that. She would suddenly and unexplainably dash back into the house in a frenzy. Attempted excursions to the movies, to church or to *anywhere* only served, it seemed, to re-enforce the fact that she was never able to get there. She began to take out her frustrations on her brother. She couldn't understand why she was having so many problems and he wasn't. He, in turn, was trying as best he could to function in a household that was highly disrupted, to say the least. Because we did not even know, ourselves, what was going on, we could not offer any explanations to him. (I have nothing but admiration for this young man, who, while all of this was going on, began his very first year in the same junior high school of which we speak.) But, true to his own strong spirit, he trodded firmly on his way, to grow and to become the wonderful young man that he is.

As long as Karen was unable to get to school, I felt I had no choice but to remain in the "counseling" sessions. Two months went by and nothing had changed. I remember mentioning to someone at the time that we probably could have been the first test case in the state to obtain home tutoring for a "gifted" child. Little did I know! *Finally,* after many months, they did request a home tutor to be implemented as part of the treatment plan. Too bad it wasn't instituted with the dignity and honor with which it should have been.

I continued to go to the weekly "therapy" sessions with the same female "therapist". I hated to go to these sessions. To me, they were such a waste of time and money. Karen went to the facility weekly to see the young female student therapist. She went without a problem now, but always maintained that she was not talking with her. She said that all that she did was sit and color. She would go as long as that's all she had to do and if going meant continuation of home tutoring.

Her real focus of communication and comradery, though, was with the young caseworker from DCYF. They had so much in common and were really becoming good friends. I was thrilled that Karen had a female friend, other than me, to whom she could relate. I was elated as I watched their friendship grow. Perhaps she could serve as the one to whom Karen could turn to as a peer, as she needed to turn away from me, as mother.

The caseworker never lost sight of her objective to return Karen to school. Although she kept some confidentialities of their relationship, she, Karen and I communicated very well and worked together toward the extinction of some of Karen's fears. She came to the house several times a week at various hours, and for various lengths of time. The goal was to build an increasingly broad base of support which would lead Karen to go out socially and then to school. The three of us kept our relationships open as together we worked to discover and solve the problem. I remember my daughter saying to me, (with ever so much faith) "Mom, we don't need them. You, me and Karen can work it out."

Some Real "Clues"

And the three of us did try to "work it out". Karen (of DCYF) was beginning to work with me on getting young Karen back on her dietary regime. And Karen, when not exhausted from fighting her own fears, was getting some much-needed rest. We finally began to see little 'breakthroughs."

It was time for Karen to have her braces removed. We could not get her to go to the first appointment, so we had to reschedule it, hoping she would go to the second one. Just prior to the second appointment, we found a large, plump, pink, smiling round-faced bear in a nearby drug store. It looked so full of love, and was a perfect addition for the support base and for her much-loved stuffed animal collection, which she had started to build up again. "Braces", as she named him, was a real boost in helping her to get to the orthodontist.

Karen and I accompanied young Karen to the orthodontist and it was there that the window to enlightenment was opened--ever so little--but even a little was a lot to us.

The orthodontist, at the every outset of the visit, told Karen that he was going to remove her braces and then she would be more beautiful than ever. He made "small talk" about boy-friends and the all the other trite remarks that usually interest teens. Karen (the staff worker) and I thought little of those casual remarks, but we both sensed some sort of reaction being "triggered off" in Karen. We began to get the first clues that something seemed to be "surfacing".

Karen had always been a beautiful child, graceful, neatly groomed, petite, a "typical" ballerina. What meaning was the orthodontist's statement having for her? We did not know, but we knew it meant something. When she got home she ran to her bed. Lying down, she began to sob uncontrollably. We stood helplessly by as we heard what was the beginning of the "unlocking" of those pent-up emotions: "It's not good to be pretty." "It's not good to be smart." "It's not good to get good grades." We knew something was happening--and it had to do with identity in a world full of teenagers.

The older Karen and I, ever present, stood by while she "blurted" it all out. She told of how she seemed to be so different. Everyone at school seemed to be involved in activities different from those that she was doing. Everyone now was baby-sitting. She was not. Everyone was concerned with boys. She was not. She loved ballet. Everyone else did not. She talked about her love of ballet. She said she was not doing it "right". She had read books and had learned that to do it "right" your feet were supposed to hurt. Hers didn't. "Real dancers" were supposed to be dancing and practicing so much more than she was. She knew hers wasn't "right" for the way serious dancers are made. I was amazed at this outpouring. I knew she loved ballet with a passion, but that much? Oh my God! What's a mother to do?

We began where we could!

She had mentioned baby-sitting in that outpouring of emotions. Maybe we could begin with that issue. I had heard that a baby-sitting program for teenagers was going to be given in another community. Perhaps this would help her to identify somewhat with her peers. We decided that she would enroll in the next eight week course. When the date of the course arrived, she and I drove to a city nearby where the class was to be held in a high-school there. When we arrived, there was trouble again. We managed to get into the school building, but when she saw the group of young girls in the classroom, she would not enter. She became panic-stricken, ran into the bathroom and locked herself in one of the cubicles, all in what seemed like one split second. One minute we were walking into class, the next minute she was locked in a bathroom in an unfamiliar building. What was this "thing" that was "gripping" her emotions? Once more, the "powers that be" afforded me patience and presence of mind and after a long period of time, I was able to get her to come out from that locked bathroom. Needless to say, we went hurriedly home. As I pondered all this, I could not help but wonder, "what else was somehow 'locked' into her mind"? What messages had she received implicitly or explicitly that was making her so afraid? These behaviors were not "games". One needed only to be there to witness the fear, the anguish, and the "panic" that went with them.

I would not give up. We *had* gone into the building and gotten as far as the classroom. In terms of the over-all perspective, this was real progress. The following week, we made another attempt to attend the course. Although she did not manage to get as far as the classroom, she was at least able to go to the school, and to stay in the cafeteria for awhile.

By our standards, she was making remarkable progress. She was in a school environment, in a cafeteria, no less, where there were other people! By the following session, she was able to sit in the classroom for a short time, while I waited for her in the cafeteria. Little by little we were making progress, and, believe it or not, she was successful in getting her baby-sitting certificate!

If it seems to the reader that I was going to enormous lengths for my daughter, you have to believe that I did. There is no question in my mind that I was doing the right thing. I am a mother. What if they had used their judgments to make a case against me because she was out of school for so long without a "medical" diagnosis? (as I was later told could have happened). What if they had succeeded in taking her from me, as I have heard has happened to others? Where would she be? What would they have done with her? I do not even like to speculate. But now that I can, I believe that we all need to ask ourselves some serious questions about "expert knowledge" and the value and sensibility of some kinds of "authority" and "rules".

[We all share in this problem. Where would this bright, gifted girl be today?...a school drop-out at age 13? Or worse still, left in whatever state the increasing depression would lead her? We all have to ask ourselves these questions, for if we are not part of the solutions, then we are part of the problem in letting it continue. "What part do I play in fostering or preventing these kinds of situations in society? What kind of a society do we have where ignorance and/or hollow "authority" rules?]

Something unknown was "gripping" my bright daughter. Months had gone by and she still had not gone to school. Her friends from school continued to call and express interest, but they, just as we, could not fully comprehend what was going on. Although it was becoming clear to me throughout this, that her immune system was just not functioning properly, we really did not know what to tell people. We encouraged continued contacts, as we kept our focus on the attainment of our goals, with the ultimate one of returning to school.

My husband was *very* concerned about all of this. Where I had always been a person to take action in attempts to alleviate problems, he tended to worry enormously. He showed much emotion because some "thing", that he did not understand, was happening to his beloved daughter. I think that what held me together at least initially, was my knowledge of adolescent behaviors and of how disruptive hormonal systems can become during puberty. I also had a lot of faith. Although I could not really define the part that each of these was playing, I could understand how the addition of pubertal factors could enormously tax an already deficient immune system. But I also felt that I needed more knowledge about this "thing" called "gifted".

I joined an organization which was relatively new in the state, formed for the purpose of advocating for gifted children. I attended some lectures given by a clinical psychologist who worked with "gifted" children and discovered that "giftedness" can manifest itself in different ways, that not all "gifted" children learn alike or are high academic achievers. I was on a path to further knowledge, but didn't know or realize until much later, that "gifted" went beyond even her understanding of the term. It includes the much broader aspect of the "creatively gifted".

By now, the two Karens and I were working together beautifully as a team. We managed to get young Karen to occasionally go to nearby restaurants for treats, something she had not done since last summer at Martha's Vineyard. Now that she was rested and more reasonable, I was able to talk to her at length about her environmental and food sensitivities. As she was gaining in maturity, wisdom and experience, her condition was beginning to make sense to her. This was an important "beginning".

It was now January. Home tutoring had been initiated in November and she was still out of school! During that time, we had remained in close contact with the guidance counselor at the junior high. We were always working together, looking for an opportunity which might get Karen back to school. She was doing remarkably well scholastically, not only catching up to her classmates' pace but also getting all A's! She was happy with the home tutoring and thankful for the opportunity to learn. We watched the depression lifting.

We still attended the weekly "therapy" sessions so that the tutoring might continue, but counseling, as I know it, was not happening. The therapist whom I was seeing did finally suggest, though, that we seek a school where Karen might be able to meet her full potential because she was "very bright".

My search began. I did not care what cost there might be. Regardless of cost, we needed to get her back into school. I investigated many private schools, but along with her "brightness", I also had to tell them the rest of the problem. She was also--for some strange reason-- "housebound". How could we get her to attend any school when our only success in getting her out was for an occasional snack and once to a baby-sitting course? But I needed to try, as hopeless as it might seem. I went to the few private schools in the area, tearfully telling my story each time, asking if they would accept Karen as a part-time student (if we could get her to go). How foolish I must have sounded! Although they all sympathized with the problems we were having, non could accommodate those arrangements. Besides, Karen had said that she wanted to go back to the same public school. She loved being in the band, playing the clarinet, and she was doing well in the accelerated program. She had said that she wanted to continue at that school. The problem remained, then, how to get her there.

It was not easy for me to be continuing my own professional development as a candidate for a Master's Degree program with all of this going on, but I was. Along with that I had secured a position of volunteer counseling at an agency. This coincided nicely with the courses that I was taking in Agency Counseling and, somehow, I was able to keep my objectivity even with all of the problems that we were experiencing at home.

Karen with the help of the DCYF worker returned to her health program of vitamins, minerals, amino acids, yogurt, small frequent meals, rest and other precautions. Changes began to occur. She was no longer terrified to be left alone in fear of the return of the frightening "attacks". Even though she was still mostly homebound, she was experiencing a period of relatively good health, back on the diet and getting plenty of sleep and rest. The home tutoring was going very well. There was close continuity with the school and the regular classroom teachers and Karen's enthusiasm for learning was intense again. The home instructors were amazed at the speed with which she grasped things and the creativity which she always demonstrated. One of her home teachers, long involved in teaching students in accelerated programs, made a special notation for the school system of what he saw as "outstanding capabilities."

One teacher with whom I had maintained a particularly close connection, was the band teacher. Karen had so much loved being in the band and many of her friends were in the band. I thought that this might be the link to pursue if we were ever going to reconnect her to the school building. In order to keep her current with what the band members were learning, I arranged for private lessons at the home of a private instructor, who often worked collaboratively with the music teacher. I continued with my own goal of trying to keeping Karen aware of and involved in all that was going on in her school classes so that, at any given time, she could realign herself with her classmates. But even going to the music lessons was not an easy task. Only after some initial trepidation and false starts, was I finally able to leave her at the music instructor's home alone for the half-hour lesson. Once again, she learned quickly and well, but was still experiencing the unexplained "fear", although it was nowhere near as problematic. For one thing, we were better able to cope with it and knew that it had to be a fact of our lives right now;

Karen was now receiving the benefit of her school education, as she should have been. The guidance counselor, (still very much involved and genuinely interested), was giving her supplemental assignments and making sure that her friends were contacting her often. We were all trying. Karen was able to get the appropriate rest that she had needed for so long and she was back on her allergy treatment regime. She began to improve ever so much in her looks, energy level, and well-being. Her appetite returned and she ate nothing but home-cooked, natural foods again. I treated her as though she was attending regular school. There was a scheduled time for getting up, for lunch and for "quitting". No outside activity until after school hours. She still did not wish to go outside for fear that she would meet any of her classmates and not be able to explain to them why she was not in school, (a realistic concern). The problem was not that she wasn't justified in being out of school, she just didn't know what to tell them about them about the "why" of it.

Even with her out of school, there was some semblance of

family stability now in place. I was again able to engage in activities outside of the home. I was participating in the community as a member of the local library board. This was enjoyable and stimulating for me, socially and intellectually. I was active again and contributing to the community in a productive way.

In that capacity, I happened to attend a dinner meeting one evening. A young woman arrived in haste and sat beside me. She told me that she was a reporter, there to cover the event for the state newspaper. As we spoke, I told her that I had a daughter who had a keen interest in writing. The reporter said she would like to meet her and suggested that they meet at a small, local restaurant. I did not tell her of the problem we were experiencing, that maybe Karen would not be "able" to get there. It did take two scheduled dates, but they finally met. The reporter truly enjoyed my daughter and they met a second time at my home. They discussed various colleges, particularly as they related to studies as a college Major in English. They exchanged telephone numbers and left with an open goodbye. We did not see the reporter again, but saw an increasing number of news articles, written by her, appear in our state newspaper. We were happy for her success. A few years later, one of them won an award. It appeared as a week-long series of articles on the front pages of the local newspaper. It was an autobiographical account of her battle with agoraphobia. She described her struggle and victory with some of the same strange symptoms which my daughter was now experiencing!

What is This Thing Called "Giftedness"?

I was now truly enjoying my graduate classwork and was taking a course titled "Introduction to Social and Rehabilitative Services." For my term paper I chose to research the topic of "Giftedness"!

It was now 1981. Within the last ten years there had been a slow, but steady, increase in interest in gifted children in our country. We were beginning to hear of and to read more about this particular segment of our population. Our local news media had brought the gifted child into focus in our own state in recent weeks. One of T.V.'s most popular daytime shows, "Donahue", brought to national attention, the fact that a family in our state was under a court order for keeping their gifted children at home to educate them rather than allowing them to go to school. In September, 1981 our major newspaper published an article titled, "Intellect becomes 'handicap' for some gifted pupils" and about the same time, our local T.V. station aired a half-hour special on prime time, telling of the plight of some families of gifted children in our state. It described the roadblocks and frustrations that these children and their parents experienced in their attempts to find an appropriate way of educating their children. The children of families that were interviewed had, unfortunately, not found a niche in our society's school system into which they could fit. Many of them were now being tutored at home because of this. One child had been labeled as having 'school phobia' and one set of parents were under arrest for failing to comply with thc compulsory school attendance laws of our state. Three days later, a radio news item related that three families in the northern part of the state, whose children were gifted, were about to submit a plan for approval so that the children could be educated at home. Yes, there was definitely a lot going on about the subject of "Giftedness".

The culmination of my own research seemed very much related to the core curriculum of Social and Rehabilitative Services, and my thesis substantiated reasons why counseling and understanding is needed for many, (not all), who are given this label.

The topic seemed to be shrouded in misunderstanding, lack of knowledge and sometimes non-acceptance of children who hold this characterization. Much that I had heard and read did not sound good or encouraging for many of the "gifted" children. I learned of how many of them experience social isolation and have a particularly difficult time during adolescence, if they do not have peers who match their characteristics. I found that some of the very factors which I have described as being characteristic of the gifted and/or talented and which can contribute to the development of their potential, may also, or even instead, cause frustration, misunderstandings and be generally problematic for them. It was disturbing for me to realize that the segment of our population who have the potential to offer society its greatest gifts, are the very ones who cannot seem to find their rightful place within that society. There definitely seemed to be some correlations that needed clarification. Perhaps it was not an error that the first unit that was established for the Gifted and Talented was in the U.S. Office of Education, Bureau of the Handicapped. I titled my paper "Giftedness - Does it Have to be a Disability?"

Looking back, I find that I had made, what I thought were some interesting observations. I had noted that not all of the children who had been in the gifted program with my daughter were doing as well as others, particularly those who had shown a great deal of creativity. More than one of them was showing signs of "floundering" within the system. It also seemed that some, but not all, of the <u>academically</u> talented who had not attended Project Gifted were outpacing some who had been in the program in grades, self-confidence and self-esteem. I wondered if the needs of some of the creatively "gifted" ones just were not being met by an "accelerated" program. The accelerated programs that my children were in seemed to provide challenge by way of increased difficulty and pace of content to be mastered rather than providing the atmosphere necessary for truly creative minds to flourish. As I look at it now, I believe that schools' accelerated programs are probably geared to meet certain standards and criteria for "mastering" known particular tasks and knowledge, preparatory to college, rather than to the promotion of creation or invention. (Basically, how well do you know, can repeat and "fit into" someone else's concepts?) Those whose skills seem to fall into the latter category and who demonstrate some creative skills, are often steered into vocational programs. I believe that this factor (of creative v.s. "status quo") was one of the dilemmas that we experienced. My children met the "standards for excellence" in the accelerated programs, but the creative aspects of their talents were necessarily minimized in programs that were mainly geared to meet someone else's already established standards. I still cannot help but wonder why there isn't more public knowledge and understanding about the phenomena of "gifted" children in its broadest sense, and if there is, then why isn't there a "working knowledge" of it?

As I became more familiar with the subject and the various implications that it may have for many, I became convinced that children needed the help of adults to identify them, to guide them, to encourage them and to advocate for them. I also knew that one day I would have to be a disseminator of that information, but I did not know how or when that would come about. I did know, though, that in light of my daughter's present experiences, I needed to begin to spread that knowledge, at *least* to those people who would be having anything to do with the "gifted population".

I began by sharing that paper with the school guidance

counselor, realizing that I had made some concrete findings about some of my daughter's "problems". I also let him know that I had found some comfort in the knowledge that I was not alone as a "Mom" in having children who had experienced these kinds of things.

He had guided and counseled many children for a number of years and was quite impressed by it all. He was also glad that I was pursuing a Master's degree in Agency Counseling where someday I would probably use this data more extensively. I also sent the completed essay to a state college professor who had planned the new curriculum for teachers of gifted programs. He gave it a serious review, finding it to be "informative", "offering a "new perspective" and "a fresh approach" on the subject, which he said was "needed". I did not realize until much later how relevant his comments were.

I thought it was also important to share my findings with the therapists, whom we were still seeing. Karen was still out of school. It was apparent from her remarks to the young DCYF worker and myself that were "blurted" out following the visit to the orthodontist, that there were, indeed, some identity issues. Did those therapists understand the information that I had uncovered? Were they applying that knowledge in the course of Karen's therapy?

Karen was still being seen on a weekly basis by the student intern at the facility. She was still telling me repeatedly that she was not talking to her counselor and was only drawing at the sessions there. Up until this time, I did nothing to question or interfere with what might be going on at those "sessions". As a professional and a mother, I realized the importance of respecting my daughter's privacy, especially in her teen years. Art therapy has its place. Maybe they were finding something. Far be it from me to question another professional's methods.

But what *was* happening there? Many attempts made by the DCYF worker to make an appointment for a joint conference to give input about our home environment and to coordinate home efforts with their program of "therapy" had been ignored or refused. I had not made any contacts with her counselor, but felt that now was the time to do so. I thought that, by now, I should have some sort of "progress report" on Karen's "therapy". I had not received any from the therapist that I was seeing. The whole idea had been that one therapist would work with "mother" and one therapist would work with "child" and together they would collaborate their efforts and "findings". They probably had no "findings" on which to collaborate. I also thought that I'd better share my insights about gifted children with the young student intern.

The student therapist agreed to see me. We set up an appointment to meet without my daughter. (She was personable and seemingly intelligent). She confirmed that she had not been able to get Karen to talk about anything other than superficial trivia and so, yes, there was a lot of artwork (drawing). She had "not been able to discover any real problem". "Naturally", I thought "because there is none"- outside of the ones I already mentioned and had been fervently working on. I was very angry, though, when she confirmed what Karen had told me, that she was only drawing at those "sessions". We were the "helpees", looking for help and we were the ones doing all of the work, including the professional aspects of coordination with the school. And we, (along with an HMO), were also paying for the services.

The young student intern was as astonished as I was at hearing what the real story was and what the problems might be. She had no inkling! She never even knew that Karen took ballet! Why were we there?

I sincerely wondered why we were investing the time and money. It was for no other reason than to substantiate a diagnosis that would warrant home tutoring. But, I was angry at all of this. Now I really did have a problem to discuss with my therapist. At the next visit I told her about my visit with Karen's therapist and how she had known nothing about Karen's conflict about ballet, her frustrations at her inability to attend classes, and the heart-wrenching conflicts that she continued to experience. She finally seemed to listen to me. I told her that this was a problem that we had mentioned at the very *first* visit to this facility. The problems were not being resolved here at all. With this said, and a home tutor well in place, we was able to negotiate a plan. We made an agreement whereby we were able to forego the sessions. I and the DCYF worker were having successes. The tutoring would remain in place and we would return, if needed. For now, I guess, we were "cured". I left my essay about gifted children with the young intern therapist and with the older therapist with the advice that they read it as a source of gaining knowledge.

Success Comes in Small Packages

In the meantime, Karen, (the DCYF worker) and I were still trying various strategies which might get young Karen to school. I had gone to the junior high school and spoken with the classroom teachers who were assigning the work to the tutor and receiving the reports from him. They were genuinely concerned and wondering what was going on with this bright student. Some of them knew my husband from his work at the Boys' Club and expressed their compassion that we were experiencing this bewildering problem. Some teachers that she had in the elementary programs heard about was going on. We received calls and/or visits, offering to help and support in whatever way that they could. But they, neither, could fully comprehend what was happening.

Throughout, all this, both of my children maintained high grades and scored consistently in the 97th and 98th percentiles nationally on every achievement test given from elementary through senior high school. Both were extremely interested in music and seemed to possess some aptitudes in this field, as well, as many others. In fact, Karen's association with the band seemed to be a crucial link to the school.

In my efforts to help Karen keep current with all aspects of school activities, she had continued the private clarinet lessons with the instructor recommended by the band teacher. She knew what the band was currently learning and doing. Thus, we hoped that when Karen returned to school, this would be another area in which she could just fit right in. (I never for a moment gave up this goal or doubted that it would be attained.) The band leader was truly concerned and eager to help toward this goal. He, too, was fond of this young, bright, conscientious student. He, the guidance counselor, the young caseworker, Karen, my daughter Karen and I were in constant discussion about getting Karen back into the school setting, even though it was now April and she was nearing the completion of her eighth grade school assignments at home.

"Open House" night, was approaching. One night each spring students and teachers hosted an evening of information for parents. The guidance counselor telephoned and asked Karen if she would like to be a hostess for the evening, handing out literature and directing parents to designated areas. Although she had not been attending school, she had long been in a state of much improved physical health. This, in turn, had enabled her to achieve a much better stance in coping with the emotional and psychological aspects of life. And this time, she was able to answer the call. Her highly social nature won out. She really would like to - she would try - and this time she made it! It was a very big step. Little by little, our combined efforts led to her return to school and to the beginning of the school year for her....in April.

She became so happy and enthusiastic to be part of things once again. She was happy to be reunited with her friends in the band. She went to the slumber parties which were almost weekly. I watched with caution. She needed to "get back" socially, but would the active pace bring a return of the symptoms? I needed to let nature take its course. And so, I did.

Disaster Returns

"Nature's course" allowed a return of the problems. Once again, lack of sleep, excitement, "teenage food", and the environmental allergies of the season took their toll on her. I barely wanted to admit, even to myself, that I saw that lackluster and exhaustion taking over, but they were. And so, on one morning in May, as the last weeks of school approached, I was not surprised to hear her say that she could not go to school. She did not need to tell me why. I knew. But, how many others really knew why?

When I spoke with the guidance counselor, he sounded shocked! "But, she was doing so well", he said! "I don't understand it. I saw her at a dance the other night, all 'bubbly and cheerful.' It's got to be something physical", he said. I responded, "I know, but I just can't find it".

Karen was not able to attend classes thereafter. She had been anxiously looking forward to the happy closing activities of the school year. She had bought herself an autograph book and, each evening, prepared the clothes that she thought she would wear to school the next day, something a little special, because they were the last days of school. Yet, each morning, the accumulated exhaustion was evident and, once again, she stayed home. There were not many more days left to the school year and she had successfully completed all the required grade work. Even though we were all still making a concerted effort to solve the problem, she remained out of school until the very last day. That last day of school was to be only a half-day. She felt that she could probably survive that, and if she couldn't, there would be no problem in her leaving. It was only for a morning, but she was successful. In spite of a horrendous school year, she received her usual A's and B's, mostly A's. We did not breathe a sigh of relief. A school year had ended, but it was apparent to me that her problem had not. It would not end until we had definite knowledge of all of the causative factors, their implications and how best to resolve them. We did know, however, that we had to work on those physically-related issues that had once again sapped her energy. Back on her diet was a starting point.

Her inability to attend school again, particularly at its closing, had undermined her self-confidence. Her appetite was decreasing again, but she *did* want to retain contact with her friends. After a few weeks, she went to the beach a couple of times with them but always came home exhausted. The hot sun, enjoyed by most teenagers was, for her, only a source of more exhaustion. She would have to find that out by herself and live her life accordingly. I allowed her to continue these activities as best she could, with supervision and advice from a distance. It was important for her to maintain her social life and relationships and, in time, she alone would have to find the "right" balance for her.

The occasions when she met with her friends that summer were not often, but, when they did get together, they all talked about the band with enthusiasm. Band camp was held every year during the latter part of the summer and was soon to begin. Karen often talked about it, excited about the adventure and opportunities it offered. The practices were held every morning during the last few weeks of summer just prior to the opening of the school year. They were getting ready for the season's major events, preparing to play at Disney World and elsewhere.

Although she spoke about the planned bus trips with an eagerness in her voice, her spirit did not match. When the first day of band practice came, (still during summer), she was unable to go. Although she got ready and started on her way up the street many times, she never got there. I contacted the band instructor, who had already lived through this experience with us during the past school year. We were ready to resume something of the same kind of plan that had worked before. That meant, basically, that she was just going to continue to try to go each morning. That in itself was O.K., but once again, each try that resulted in non-attendance was matched by the same amount of discouragement and depression.

The pollens, accumulated junk foods, the days at the beach in the sun, although limited, had taken their toll on her body again. It was not up to me at this time in her teen-age life to place unrealistic restrictions upon her. I could only stand by at this point and tell her what I saw happening-- to help her to see these problems in view of their relationship to those she had experienced in childhood and to tell her of the effects that allergens and environment had always had upon her system. I prayed that the day would soon come when she would "take over" these problems and resolve them in her own manner and to her satisfaction. That time had not yet arrived. Experience, as they say, would be the only and "best" teacher.

I watched her fruitless daily efforts to attend the band practices. I heard her on the telephone, talking to her friends, trying to explain to them why she was not attending and I watched depression occurring. She became more and more depressed, then began to stay in, and did not call her friends. Once more, they couldn't understand. Neither did she. (Neither did I, nor the next therapist to whom we went for help.)

Her depression increased and she reached the point where she was fearful again. During the summer months,in typical teen-age fashion, she had been off her health regime again. I was convinced now, that every time she had failed to stick to the diet and adhere to the restrictions that had been prescribed so long ago, she experienced the same kinds of problems. But how does a parent of a teenager "make" her stick to these regulations? And would I really have been wise to make her do so? I took the road that made more sense to me, but a more painful one, I believe; that of a parent who is standing by waiting for maturity and responsible adulthood to "happen". She needed to grow and to have the independence of making her own choices and to feel their consequences, however painful. She had been put in a further place of conflict by the many people who were telling her that those "allergies" could "never have caused the problems that your mother is saying they had". They were right. "Allergies" were not the sole culprit of her symptoms. But "allergies" had been the only name given to us. We found out much later that there were many more complex biological factors involved. The problems that she had experienced with sleep patterns, fatigue, hyperactivity and the others, had a more extensive and interactive physical basis. Later research would confirm that. For now, we had managed to keep her fairly well through the integrated holistic program of treatment that I had put together for her. But, we still had to face the hormonal effects of puberty and were about to experience yet one more trial which, this time, drastically affected our family unit.

Up to this point, I had done well, I think, taking each event and circumstance as it occurred, looking for the presenting problem and the corresponding answer for it. I had "kept the faith", (my own degree and type of spirituality). I did everything that I could to ensure that my own mental and physical health stayed intact. I continued taking courses in the pursuit of my Master's Degree. I maintained my bicycling and swimming. I went to plays and concerts with friends, when possible and appropriate. In short, I continued with the holistic program that I had devised for myself through the years in an effort to combat my own physical problems and move forward with my life. To be sure, my own health had greatly improved with the program of vitamin/mineral supplements, meditation, and the other "natural" therapies that I had discovered and been using. I no longer needed to take thyroid or other medication, (much to my doctor's surprise), and I was no longer having the problems that had wrought havoc with my reproductive system in earlier days. Moreover, I had not had yet developed any of the "serious" breast problems that I was told a few years earlier to expect, following the excision of a few small lumps. Bouts of illness for me had become fewer and were much shorter when they did occur. I was taking a variety of amino acids, vitamins and minerals, including the anti-oxidants, long before telling my physicians. The philosophies and practices that I had found were not yet considered part of standard, acceptable medical treatments in the circle of professionals with whom I had associated. My spiritual practices were sometimes akin to those of an organized church, but I could not follow the rules as prescribed. Nevertheless, I believed that I was a very spiritual being and I certainly had gained in "wellness".

Although I was practicing a lifetime worth of learning, Karen was still young and was not yet founded in her own health practices. She was now experiencing the effects of <u>her</u> life circumstances with which she would have to contend.

She began increasingly to withdraw after her failure to attend the band practices and the more activities she missed, the more she seemed to withdraw. I tried to plan activities that she had always enjoyed, with her cousins and others, but I was unsuccessful. "It", (whatever "it" was), was happening again. As before, she would go out, but once she got to the car or to a group, she would literally "cling" to me. One day I took her and her brother to a theater where there was a movie playing which they had both previously expressed they had wanted to see. Surely, there should be no problem with that. I took them to an early matinee thinking that there would be fewer people then. (I did not particularly want to be with them, but thought it was important to try this venture.) As the three of us started walking toward the theater, Karen turned around, darted back to the car and started to cry. She "could not go there". After that, I was not able to get her to go anywhere outside of the home again. The strange malady had returned. The words "panic disorder" or "agoraphobia" were never a thought in anybody's mind, nor were they even known or suggested to us. I don't recall of ever having heard those words at that time, and, if I had, nobody was using them now. Besides, what good is a label without a cure?

I dreaded the coming of school. It looked as if Karen would again be experiencing problems in attending, while it would be my son's first year at that school. Having finished Project Gifted, he was ready to return with his peers and had made the decision to attend that public school.

In spite of our family difficulties, he had maintained his scholastic standing and fine reputation and had been offered a scholarship to a private school. But his love of music beckoned to him. He wanted to pursue the talents that he had so clearly demonstrated from early on with his award-winning accordion playing. The city's public school had long been noted for its band program. So, even though the private school, when hearing his decision not to attend, increased the amount of the scholarship that they offered him, he stood firm in his decision. He would attend the city's public schools for its music program and would begin there in September. He would be starting in seventh grade while Karen would be beginning ninth, the final grade in that school.

The school year was fast approaching and Karen was again home-bound. I knew I had to get some help fast. I decided to contact the psychologist whom I had earlier encountered and whose specialty was counseling the gifted.

A New Therapist, A New Vision

I thought it would be wise to seek professional help again as soon as we realized that the problem had returned again during the summer. I did not know, though, what other kinds of "help" we needed or would find. Three times now, in the last few years, we had lived through prolonged periods of difficult circumstances and behaviors with Karen. They definitely seemed to be related to some physiological disturbance related to a nutritional state and environmental allergens of some sort. I did not believe that the causative factors had ever really gone away; they were just manifesting differently. The "biological" part of the problem was already well-known to me. The solution was only a matter of Karen needing to realize her own physical make-up and tending to it appropriately. But now we had the added burden of adolescence with which to contend. Hormonal chemicals were acting on her system, while she was already trying to cope with these other factors. And now it looked like there might be another issue, that of identity, which seemed to be complicated by this aspect of being "gifted". I was beginning to zero in on the issues with which we needed help. I hoped that by the beginning of school, we could get that help and that we would have things pretty much under control.

I made an appointment to meet with the therapist whose specialty was counseling the gifted, whom I had earlier heard give a talk on the subject. She had done her internship as a clinical psychologist at a nationally known children's facility and was presently a staff member in yet another facility that serviced children. She had experience in working with gifted children, individually and in groups. I thought that those credentials were relevant and sufficient and would serve our needs. It certainly did help with providing us with facts and understanding about some aspects of giftedness, but for a truly therapeutic process to have occurred, (I discovered later), that we needed a person who had actually gone through the process of raising gifted children within the intricacies of a real family-who was experientially knowledgeable about the problems of getting to little league, of saving for a college education, of large ethnic family values and of the stresses of being in a male/female relationship that began in the 60's. Without the latter criteria, she could shed some light on giftedness, but could not relate to the problems of our real-life context. In addition, she had never before seen my daughter or me. Prior to our meeting, she knew nothing of me professionally, of our family history, or of the terrible ordeal that we had been through. In fairness to her and her skills, I have to preface this portion with the statement that she truly did inherit a "mess", for, not only did we need help with the concept of giftedness, but also with this other "unknown", as yet undetermined, physical malady.

When we presented ourselves to this therapist, my daughter was again at the point where she was "gripped" by whatever it was that was possessing her. I was a distressed mother, nearing my wit's end, distraught that this was happening all over again. Any semblance that we had known of a somewhat "normal" family, with high-functioning members was quickly disintegrating and probably little in evidence at this point.

I informed the therapist of our experiences of the last two years and how Karen had "recuperated", but only for a few months. We had been through so much and now it seemed that the whole thing was happening again. She had become depressed. She cried often and literally "clung" to me when I was about to leave to go anywhere. She was terrified to think that I might go somewhere and leave her "abandoned". I could not understand this. This was not my daughter. This was not the little girl so eager to try new things and meet new people. This was not the little girl so eager to try out the experience of going to Project Gifted in a new school district, saying, so confidently, "I'll make new friends, Mommy". But this was the picture that we presented to that therapist-- a frightened, clinging young adolescent and an exhausted, discouraged mother trying to give her daughter the much-needed security that she seemed to need.

It was a complicated mess, but I hoped that this therapist might be able to shed some light on what was happening in order to help us resolve the problems and to provide ongoing advocacy with the school system in the area of giftedness, which was her stated specialty.

For awhile, the sessions went O.K. The therapist confirmed very readily that my daughter, son, and even I, she said, amply and aptly exemplified the characteristics which the word "gifted" implied. She was able to work with Karen on some areas, such as "feeling different" and other issues of general concern for gifted youth, in which she seemed so well-versed. She also provided relaxation techniques which Karen practiced and seemed to benefit from. She understood Karen's high level of sensitivities and imagination. She was able to uncover and work with some of the fears which Karen seemed to have picked up from those late night teenage talks during her camping experience, and which had seemingly become "locked" in her mind. (Teenagers in campsites are often known for the gruesome tales which are told in the darkness of night and my highly imaginative daughter very readily absorbed the impressions of some of those fearsome stories.) This therapist was doing some "good" things, initially.

But, she too, in time, went badly astray from the real issues. Well-meaning, though she was, the counseling process eventually became derailed and "out of context", not only because she was wrong, but because she was not open to that which might be right. She substituted an inaccurate perspective of our family with the mandate that we buy that perspective. Because of her insistence on her "right-ness", I was eventually left scarred and temporarily destroyed. Further havoc was wrought upon our family for the next several years. In the end, I was left to the task of healing, not only my daughter, but, myself as well, and the damage that had happened to all of us.

When school came, Karen was still not able to get herself there even with the introduction of the relaxation exercises and talks with the therapist. She was again exhibiting the same set of symptoms and behaviors with which we had coped for well over a year now, which had at times subsided, and which had all started with a simple camp experience.

Several days of the new school year went by. I again contacted the guidance counselor who had worked with us previously. I informed him of the situation and of the steps that we were taking to alleviate the problem. He expressed the utmost confidence in me, having seen me resolve the problem satisfactorily the previous year. He still expressed concerns, though. This was the second time round. And this time he told me that he had seen other young bright students with similar symptoms. He had noted that they came to school for very short periods and then "shot" out of the classroom in a panic to go home. He expressed his fear and concern that Karen might end up like one of these children, never being able to return. I remember saying to him emphatically, "that won't happen".

I needed to cooperate with the present therapist in any way that would seem to work. I gladly had Karen re-evaluated by her pediatrician. We also returned to the allergist for a complete evaluation. He again confirmed that her allergies were from many and varied causes. He also re-confirmed the symptoms of fatigue and the stomach problems as allergy-related. These were in addition to the respiratory problems that we were now seeing and that were also allergy-related. But no, there did not seem to be any readily available explanation of why the symptoms were severe or continued for such a long period of time, in spite of the treatments. But there was no change in the basic medical diagnoses. All of the symptoms were "allergy"-related.

To complicate matters, the pediatrician who was coordinating the plan of care with the allergist, and who had really understood Karen's condition, suddenly gave up his practice. There was no forwarding address or way to follow-up with him. We were in need of some other medical person to coordinate the medical findings with those of the allergist's for the psychologist and the school. It seemed easy to choose the general practitioner that my husband and I had been going to for years. He had never seen Karen and knew little of her, but for sure, he would help.

It was difficult to tell the whole story again even though he needed it for a medical history. It had all become so complicated. He listened and sent her for another round of medical tests. These included sleep analysis, mononucleosis, blood-sugar, etc. They all came back negative. There was no other answer. My daughter simply had "allergies".

Meanwhile, Karen and I were each having separate sessions with the psychologist for the gifted. During that time, the therapist seemed to have made a "diagnosis" and on which she was basing her treatment. Evidently, the exploration of the reasons why Karen was feeling "different" were being uncovered. But the aspect of "differentness" as related to "giftedness" did not seem to take priority. The thought that seemed to prevail, was that I was the "villain" who would not let her teenager daughter "rebel". And through a multitude of sessions, she found reasons why that was true: In addition to a family value system that was faulty and problematic, I was "too altruistic" and "too loving". I gave too much. Allowances were too little by the present standards. I treated my children somewhat differently than others. I was more restrictive. (That was so. I definitely did not like the kind of behaviors that I was beginning to see some teenagers engaging in, but I sure was no dude.)

I always played classical music on the radio; (supposedly not allowing her to hear the rock music that most teens were listening to. Funny, my son was listening to rock music every day in his room.) I had not bought my children the newly popular Atari set and many other things that were currently in teenagers' possessions. In addition, I was probably one of few parents who insisted on not doing my children's chores, who wanted them to get into the habit of typing their own school papers, to do their own laundry and many other things. (I had done many things with them and for them earlier in life, but now I was teaching them and expecting them to increasingly carry out their various responsibilities.) But the most inaccurate of all implications was that, I lived my life through my children. Apparently dismissed was the fact that I was on the board of trustees of the local library board; I was continuing courses toward a Master's degree; I was working or volunteering on a part-time basis in a field that corresponded to my course of study; I belonged to a German chorus, (and learning the words to their songs was a wonderful challenge to me;) I was in touch with former nursing school colleagues, active on a church committee. I was also one of "those women" who had baked cookies, had fresh flowers on the table, helped sick neighbors and often volunteered my professional skills. That was my chosen role. I had been a full-time home-maker. But, now I was looking forward to the time when my children would be grown so that I could re-enter my professional career and balance it with all of the "fun" things that I had built up for myself. The roles were needing to be changed and I was looking forward to that. I certainly, was in no way, preventing either of my children from developing into their "own" person.

The factors just described were my parental values--good ones, I thought. If my daughter had trouble with those, then she should have help in learning to cope with those realities and to negotiate about them, not taught how to manipulate them to her liking. Anyway, there was something going on with my daughter that had nothing to do with these issues. There was something completely different happening, I thought, and nobody was realizing or were even being open to the possibility that it could be something other than family problems and/or faulty parenting.

The therapist/psychologist, knowing that Karen was becoming increasingly depressed about her condition and absence from school decided to schedule us for a visit to a psychiatrist to determine if she had a "clinical" depression which might warrant medication. The psychiatrist asked a few direct questions at the very outset of the visit. One of the first questions, (upon learning that Karen had been hyperactive as a child,) was "did she sit and cuddle with you as an infant"? "No, she was very difficult to get to calm down. I had tried to cuddle her, but she was always squirmy and difficult to get to sleep", (as noted in her early history). Another of the questions was, "who yells in the family?" there were a few more pointed questions after which he seemed to have formed his conclusion. Karen was not "clinically depressed". His questions confirmed the earlier diagnoses. There were definitely "family problems."

The treatment plan was again aimed at treating a "parental" problem. I could not believe that it was being implied that I was in any way responsible for this terrible situation in which I found myself. It was not a "parental" problem that had caused these problems following the camp experience. These strange attacks were <u>not</u> a "parental" problem. But, "professionals" who did not know me were trying to force a false identity on me which made the "treatment", in effect, much worse than any part of the problem that we had heretofore experienced! Our once-prized family values soon became a thing of the past as we succumbed to "treatment" which might get Karen to school. Younger generations of professional in the "me generation" were able to dictate my value system in the name of "treatment"? The more that I tried to intervene, the more I was called "controlling". I couldn't believe this was happening. I had already been told that we could be taken to court if Karen did not get to school. Without the data that I now have, I had no recourse.

In a few weeks school resumed. Once more, my daughter was not there. After discussion with the guidance counselor, we decided to try to work things out as we had done the previous year. But now we had other kinds of intervention and I felt real pressure. Before, I had only tried to move us all toward the goal of resolving the issues -- getting her to school and carrying out some fairly "normal" activities in between. But the idea of a possible court order now was quite a different thing with which to cope. It wasn't as though I had done nothing about this problem already. I was extremely distraught about it and now here I was, being accused as part of the causative agent. Again, I couldn't believe this. I could not understand why so many wrong assumptions were being made. To me there was so much evidence of a physical basis for Karen's problems. Even this therapist had said that Karen had "hidden" disabilities". If she suspected that, then why weren't they being pursued more vigorously and my parenting left alone? And where would one start looking for more thorough answers to "hidden disabilities" than what we had already done? I did not know. I also wondered why no one seemed to place the same emphasis on my observations and input of how nutritional factors had seemed to be so important and how my own medical history had been wrought with a multitude of metabolic disorders. I had always believed that the problems were primarily physically based and any psychological component was due to the teenage identity factors. Although we had marital issues, so did everyone else that I knew, but these were not the primary problems. But no one seemed to believe that. It seemed to be important to the professionals to get my particular type of parenting out of Karen's way so that she could go on with the task of adolescent rebellion. This was not the issue and I knew it. How far they were from an accurate diagnosis! But I did not have a better one. There was nowhere else to turn, it seemed. At this point I really felt that I had to give up to the "powers that be".

Karen was to be ushered out of the house each morning with

the school guidance counselor watching for her arrival, as we had successfully done the previous year, but this time the therapists "treatment" stipulated that she was not to return to the house once she had left it in the morning. She was to get up, prepare for school each morning as though she were going to attend, go out of the house and go as far as she could in the direction of school. She was not to return into the house for any reason until the end of the school day. If she did return from school, she was to remain in the yard *all* day, rain or shine! There was to be no coming in for any reason. Having to go to the bathroom did not constitute a valid reason. This "treatment" was to stay in place until she went to school on a regular basis. And I had to sign a contract that I would adhere to the "treatment". I was shocked. This was not fair to a child who was not "playing games". This was not fair to me nor was it even humane. Even the guidance counselor was puzzled at this turn of events, and "new plan" but he also believed this time that he needed to conform to the rules of the system. Although he told me that he thought that I had more insight into the problem than any "counselor", he needed to comply with existing regulations.

As time went on, the therapist seemed to get increasingly irritated with my reports of Karen's failed attempts, her resultant outbursts at her inability to go to school and my own frustrations. She saw a child frequently regressing to the security of her mother. Over time, I was again placed in the role as an "overprotective mother", but with an added dimension now, of being a passive victim of a verbally abusive husband, (because, in his frustration and lack of knowledge, he was yelling that Karen please go to school). In that context, our family definitely had to be "treated", while the therapist would basically take control in the treatment of Karen and her problems.

Painfully, I kept up the plan. I sent her off to school in the morning each day with the hope of making some progress. She never got farther than the backyard. According to the contract,

I did not allow her back into the house for any reason. This regime certainly would prove if the problem were physical or behavioral. But, I soon became tormented. There were knocks on the door, bangs on the door and windows, crying, yelling, sobbing. All I could think was, "this isn't right." What is happening? Why do I have to give up control of her? This is my daughter and I am her mother who is supposed to help her, especially when she is afraid and in difficulty. I should be her advocate, not her tormenter.

I don't think there is any need to further explain my feelings to any mother who might be reading this. To keep her out was torment for me, but to bring her in was to risk having her taken from me. She and I continued to see this therapist once a week. I did not know where else to go. This was the third agency that worked with children and to which we had gone for help. We did not yet have many of the newer parental networks that have since sprung up nor did we have correct diagnoses.

The more anxious I became at the seemingly inhumane treatment of keeping my daughter locked outside, the more I was looked upon as the protective and interfering parent that they already were sure that I was. Because of my now frequent, frantic telephone calls to this therapist, telling her that I could not carry out this "treatment" with my child knocking and banging at my door, pleading with me to come in, the contract was amended to include that "the mother was not to telephone the office." I was told that the treatment was to be coordinated between the school and the therapist, and that I was excluded. The contract stated that my daughter would attend school for certain specified hours, which would be increased as the days went by. Each successful attendance for the designated time periods would result in my giving Karen some reward, such as an article of clothing or the like. This was not *my* idea of parenting. I had to obey "orders" that I would not call the therapist's office any longer. And if we were not successful soon, the next step would be that my daughter was to be "dragged" to school.

I don't believe that there are many who would believe that this could happen in this day and age. But I could not explain these strange attacks and as yet, unexplained behaviors.

I thank God for the divine intervention that was provided for my daughter. Once she found that I would not bring her in under any circumstances, she sat out there day after day, rain or shine, except on weekends. She eventually seemed to reconcile that she would be living outside for some designated period of time.

As time went on, she adhered to the "rules" of staying outside. She, like me, knew that we had to conform to those "higher powers" if we were ever to get her through this, but deep down, I believe, she knew and trusted her mom who, at this point, was letting it go to the hands of the real "Higher Power".

She watched the birds and began to play self-made games. On rainy days she sat under the picnic table. She converted her love of nature and creativity to writing poems. I thanked God for the integrity of her spirit that was beginning to return. But, after many days of this, the truant officer came. The "normal" steps for delinquents were being carried out. This served no purpose but to send an already-frightened little girl hiding for cover. Fearful that the next step would be that dreadful "court order", I seemed to have no alternative but to obey the "treatment". Next on the agenda was to drag her to school. My heart broke as I literally "dragged" my daughter to school, amid her physical protests, and obviously noted by any neighbors who happened to be home during those hours. Any form of dignity in myself, my pride in my parenting and my self-esteem was eroding. There was nothing for me to do now, it seemed.

I was angry at my husband. Why was he a person that yelled? Why didn't he handle his sensitivities some other way? This whole thing was getting worse and our marriage and family was getting farther and farther away from all that we had known. I became greatly distressed about the unreal mechanisms that I was forced to use, but did everything possible to maintain my own mental and physical health. I remained faithful to my health maintenance program, and continued the courses toward my Master's Degree. Throughout, I truly wondered about the strange rules of our society.

The Beginning of a "Miracle"

The plan was not working as specified by the psychologist. The problem persisted. Fall was bringing colder weather, and my daughter was still spending her days in the backyard. Could they really take her away from me? I couldn't bear this thought any more than I could bear the current pain of leaving her outside everyday, locked out of the house. Up to this time, I had observed the "contract" of non-interference. I had kept out of it, observing the terms of the contract, but nothing had changed. Even with the prospect of "rewards" for attending school, she did not attend. With this in mind, I decided to take on the task again, myself. It had worked the last time. There was no more to be lost and everything to be gained!

The therapist's contract called for her to be in school at a certain *time* of the day, but I tried to arrange it for when her energy level seemed to be highest. That made sense to me. I noted that this did seem to relate somehow to her nutritional status, even though routine blood sugar testing had seemed to rule out classical hypoglycemic (low blood-sugar--tests for which we now know are not always accurate.)

I conferred with the same guidance counselor with whom I had worked so closely and so long in realizing the success that we had the previous year. He agreed to work with me again, saying, "I think you know more than the therapist." I said, "I know", but I also knew that I couldn't convince the rest of the professionals of that. My sanity and a threat of my daughter being taken from me was at stake. There was no more to risk.

The plan was to telephone the guidance counselor each day as I was trying to get Karen in to school. When I thought there was a possibility that she might arrive, I would again telephone him. If she arrived, it would be "O.K." with him even if she stayed for only 15 minutes. (The original plan as contracted by the therapist was that once in school, she was not, under any circumstances, to leave before the designated hourly time span. My aim, on the other hand, was pure and simple...to get her to school...in any way...for *any* amount of time.) The guidance counselor agreed to watch for her and would inform the teacher whose class corresponded to that time to let her in for whatever amount of time she might be able to attend stay. I don't know how many calls I made to him which were followed by a "no-show". But neither one of us gave up. Day after day, we used this routine. Finally, the day came when she and I were successful in getting her to the school parking lot. All of the classes were already in session, so the element of her being around a lot children was not present. We were able to just sit quietly in that environment. Day after day, instead of leaving her in the backyard, locked out of the house, I took her to the deserted parking lot and we just sat. She was not able to summon up whatever it took to get her in there. Day after day, I asked her to go in, but she seemed unable. While we sat, we talked about how her dietary regime had helped her so often. I asked her to *please* think about that, and she did. Her trust in me was taking hold again. Day after day, I went to that parking lot, hoping and praying that she would, at least, enter the building--and she did not. Day after day, for many weeks, I could be found at various times of the day sitting in the parking lot with my daughter--never giving up. And one day, she did go in to school. She stayed for a very brief time. That was alright with me. She had gone to school. The next day, she went in again. I told her that she could leave at any time, but to try to stay as long as she could. I assured her that I would be right there in the parking lot. The guidance counselor continued his confidence in me and kept his part of the plan. And little, by little, we saw it beginning to work. Day after day, I went to that parking lot, carrying out the same routine for many weeks. She began to stay for increasing amounts of time; an hour; an hour and a half and I was always there in that parking lot--waiting. I kept my word.

Finally, we were able to negotiate for her to go in and I would go back home. When she wanted or needed to leave, she would go to the guidance counselor. He or she, herself, would telephone me and I would come. As long as she had the freedom to leave and the trust in us, it worked. A far cry from "forcing" her to go at a specific time of day, staying for a "forced" designated period of time, refusing to let her leave and then giving her a reward for "obeying" the orders! I gave her my own form of rewards now and again, as I thought she deserved them, but we did not tell the therapist how we were getting her there. I knew that she would not like the idea that I had taken "control".

I have nothing but the highest commendation for that guidance counselor and the outstanding efforts that he demonstrated on behalf of one student, (ours), when his responsibilities covered so many, and I was thrilled to attend a ceremony the following year which publicly recognized him as the outstanding and valuable person that he is!

The "powers that be" provided Karen with a greater reward than any article of clothing or cassette that I might have given her. She was appointed a special "guardian angel" in the form of a stray tabby cat that appeared in our yard. I don't know where he came from or anything about him, except that shortly after Karen returned to school on a full-time basis, I saw him sitting in our backyard on a large pile of autumn leaves. He was beautiful in the sunshine with his little orange face peeking out, perfectly blended with the leaves. I smiled when I saw him there that very first day while she was in school, but my joy was nothing compared to the love that would eventually transpire between those two precious beings. He came daily, arriving just before school got out. From whence he came, I do not know, but he was there every day. I was hesitant to adopt him, not knowing anything about him, but it seemed that he was to be ours to keep. One day in October, when I returned from a visit, the children told me that while I was out, that "somehow, the cat got into the house". "It just "ran in" they said, as one of them was going out the door. So they "had to give it some milk". It never left our yard again. I called the animal shelter and we watched the lost and found ads for pets, even putting in an ad about this "lost" beautiful kitten, to make sure that he did not already belong to someone else. There was never any response. To this day, some ten years later, he holds a special place in our home and our hearts.

Life was beginning for Karen (and for us) once again. Just as she successfully returned to school, she successfully returned to dancing! We had continued her dancing lessons on a private basis just as we had done with the clarinet lessons, until the time came when she could return to the group. Now she was able to do that, too. After a very long absence, she returned to dancing with the rest of the group!

It seems that whatever thoughts or ideas that had become locked into Karen's mind during that period of multiple stresses, were being dissolved. We did find that she had worked through some of the effects of the "scary" tales that were told in the night at camp. She was also able to resolve some of the teenage identity issues enough to believe that it was "O.K." be smart, and it was "O.K." to be pretty, even if others were going on a different path.

During the "desensitization" period, I was able to talk to Karen about what had transpired in the past several months. I was able to point out to her again how I had observed the various physical symptoms and how closely they had resembled the ones she had when young, how I had found the vitamin/mineral treatment to work and how important it seemed that she maintain this treatment. Unfortunately, she had yet to learn to make the difficult choices of developing a pattern to cope with her physical realities.

The professionals working with us, for the most part, had been denying most of the physical aspects of her condition, thereby delaying the time when she would still have to learn to cope with them. In my mind they were fostering denial of her reality, which was only going to make the adjustment process all the more difficult later. And, indeed, it would not be until later on, when out from under the "controls" of these therapy encounters, that I could help her to learn how to fit the realities of her physical difficulties into a wholesome, productive life-style. The skilled parenting, which I had practiced in the past and which had brought my children so far in their endeavors, was being badly fragmented during the crucial years, but I hoped with all my heart that I had already given both children a strong enough base to survive this crisis and to continue their growth.

My son was "hanging in there", doing "O.K.", it seemed. In spite of all that I report here he was steadily maintaining a national rank of 98% on every achievement test that was given. He managed to maintain good report cards, even though they were not always A's and B's. He seemed to prefer other ways of learning and other topics to learn. He could often be found reading the entire encyclopedia, instead of just selecting the information needed for a particular homework project. Karen often asked about him and why he wasn't having the same problems that she was experiencing. He did not have her physical pre-dispositions and sensitivities. He allowed his intensity for learning to take form in other ways. He was just, (plain and simple), different from her. My thinking (which is purely my own), is that Karen seemed to want her motivation and abilities to be acknowledged appropriately within the existing system, whereas, he really was not concerned about trying to fit into prescribed ways of learning and resulting scores,(or, indeed, any "systems"). (Later on, in high school years, I believe that traditional beliefs about the value of grades, college goals, the extent of the knowledge bases of individual teachers and their attitudes toward creativity influenced both children, although in different ways.)

But my son had made the "right" choice for him, it seemed, having passed up the scholarship to the private school, in favor of a music career. Eventually he gave up the accordion in favor of the trombone. He played in the Wind Ensemble, the Jazz Band, the Marching Band and whatever else that involved music!

Karen succeeded in completing her final year at the junior high. We were proud parents joining the others on that graduation day. We knew that we all deserved a special star for our accomplishments, but we were not so sure that this was the end of our problems.

There had been a recurrence once. And she was still in the throes of puberty. What if it happened again? (God knows we prayed that it would not). But, with that possibility in mind, I remained in contact with all of the professionals who had been seeing Karen during the past year. Although they did not seem to have the most appropriate solutions, we were still in need of them. If there were problems again, the school would know that we were "covered" by "professional treatment." (What a sad commentary on our society)! Staying with these same professionals would eliminate a process of having to start all over again with yet another set of professionals, if problems did recur.

It was enough that she would be having a different guidance counselor and a whole new set of teachers at the high school. But, I was almost glad that she would be going to a different school. We needed to make a fresh start. Karen was older and more understanding now.

She was enjoying herself and looking forward to going to senior high. She was on no medication, nor had she been throughout all of this. She was, however, on her program of vitamins, minerals, amino-acid supplements, daily yogurt, all-natural foods without additives, exercise and much rest. She was ready for senior high!

On to Senior High
A Reprieve for Karen, But Attacks on Mom

The therapist was aware that Karen was making a successful recovery but still seemed to hold to her original view that the problem seemed to be one of parent-child separation, often associated with young children or with children who have been known to feel a need to stay at home to protect a parent when all is not "O.K."). I knew that our problem was due to neither of these factors but the therapist continued to inform me, in a definitive manner, that our daughter would have a *big* problem when it came time to attend the senior high in the fall (encouraging, indeed)! I did not believe so. Karen had seemed to have learned that she *needed* to follow the strict health program and this summer, she was finally following it.

I did not believe that she would experience any problem going to the senior high on the first day. She was eagerly looking forward to it. I was now seeing a more mature woman who had come to appreciate her scholastic abilities, and who was eager to get on with her life.

Karen's days at the high school went extremely well. She loved it! She liked the structure and arrangements of the class schedules, allowing for more "freedom". She knew the importance of her diet and rest. She made friends again. She became busy maintaining her grades and pursuing classical ballet, vigorously. In January, she interviewed for and was accepted as a member of the State Ballet. This involved many hours of practice and sometimes her knees would hurt, but she delighted in her love of dance. The family had a reprieve from the terrible turmoil of the past few years.

I still saw the therapist because I believed that it was necessary to continue with some sort of "supervision" until we were sure that Karen was "O.K." within this new school. If more problems were to occur, at least I would be "covered" in the eyes of the school system. In addition, I had a need to try to have this therapist become more equally interactive with me--to try to get her to decrease her authoritative stance and to become more open to the facts, as I saw them. I thought that it was important that we share our views and, perhaps, together, gain more knowledge and understanding of the term, "giftedness". I thought that this goal might be accomplished while we experienced a reprieve from Karen's needs and problems. But such was not the case. The therapist was positive that she was the "authority" who would tell me what to do. But there really was no "authority" about these issues--just a lot of people with a lot of knowledge that was not being shared appropriately. This kind of attitude does not belong in any kind of adult interaction, I believe, much less, in a supposedly therapeutic one. But, attempts that I made to equalize the interactions were termed as "controlling".

I took a subordinate role, realizing that this was not the time for arguments and I did not need any more labels. I got them anyway. Although she would later make the statements, "they don't know everything about 'gifted'," "I wish I had taken the time to understand you", and (even), "I made a lot of mistakes", many unfounded negative words, comments and accusations were continually directed at me. I wondered why the possibility of "mistakes" had not been considered right from the beginning and what is it that makes people think that titles of authority mean absoluteness and exclusive control that they must exert over other people? I had pondered that question many times, but never more so than then.

Whatever her reasons, that therapist was not able to think soon enough that she may be in error as the absolute "authority" on the subject. Not only was she in error, but as we interacted,

I began to realize that some of the characteristics which she was criticizing in me were the very ones that have been associated with gifted persons-- the high degree of tolerance that I always had for people's differences, including the differences between my husband and I. (We often did not get along, but we had managed until now.) The choral competitions that I participated in were also criticized. I was told that I was "performing" (with the implication that I was doing so for attention). The reality was that I was performing because I loved it. Singing and the socialization with the German choral members provided balance and new dimensions to my life.

Another fault, that I was "too altruistic"; that was part of the problem that needed to be "treated". (I very often sent cards or gave tiny gifts that told people that I thought of them, little "trinkets" that I'd pick up from vacation, flowers from my garden, baked goods, a meal when someone was sick.) I thought I was "spreading cheer", but this was seen as being "too loving". I was told to give gifts only on birthdays, holidays, etc.

And there was more. I loved to have fun. I had always had a high sense of creativity. I enjoyed teddy bears, stockings at Christmas, and Easter cakes in the shape of a bunny. I had loved making things in themes for the children's birthday parties-- favors made out of Popsicle sticks in the shape of the Flintstone car and filled with candy; the Popeye game for Ray's party, (replacing Pin the Tail on the Donkey)-where the children put the pipe in Popeye's mouth while blindfolded--and others. I looked forward to my own 50th Birthday, coming up soon, secretly hoping that someone would give me an "over-the-hill" party. Those who knew me always delighted at my love of the simple joys of life, my sense of adventure and my ever-readiness to try new things. The things that were viewed in the therapist's eyes, were viewed as "immaturity". In my eyes, my love and actions were the manifestation of a most "healthy child", the aspects of the innocence of childhood that we all possess and which sometimes get contaminated. In her eyes they were to be "treated".

I also, supposedly did not teach my children a sense of responsibility!

Throughout the course of this "therapy", so many of the values that I had cherished were misinterpreted or cast down, and even temporarily destroyed. Instead of being viewed as a responsible, civic-minded adult, I was looked upon as an immature woman who needed to be "treated". But the most devastating consequence stemmed from the therapist's very strong ideas about what constituted a suitable female role model.

The psychologist/therapist, strong in her desire to impose her own viewpoint, did not seem to realize where I was in my life with the development of my career and as a woman. Nor did she seem to be able to accurately identify my role in my marital relationship, the long-term financial planning involved in meeting college expenses, and the fact that I was also still a parent to a gifted young male (as well as to Karen), and that I still had responsibilities for bringing him into adulthood. All the emphasis seemed only to be that two females, (my daughter and I), should be out there using our gifts and "pulling our weight" (so to speak),as females in a heretofore masculine-dominated world,(as the saying goes"). But I never lost sight of the fact that woman's role is also to heal and nurture. And so, one woman's ideas of what another woman should be doing, "got in the way" of relationship. Words were hurtled with force and intimidation.

As we had experienced problems within and without our family, I came to realize that perhaps some of what happened was not due solely to Karen's biochemical make-up. I knew that Karen's identity problem was probably also influenced by my current role as a female caught up in a rapidly changing society. Families were experiencing a change in structure, definition and values. There was a drastic increase in single parenting and in two parents working outside of the home to obtain increased family income. But I was where I happened to be in the midst of the changing female roles and that was O.K.

We had, early on, realized the sizeable income that I could have made returning to nursing, but instead I opted to be at home, to fulfill what I considered my role of nurturing my two children, raising them according to our value system and encouraging them to develop their gifts and talents. I never would have had the time and energy to do all the particular things that I did if I had worked outside of the home all day. Instead, we made do with the income which we had, while I fulfilled my personal requirements for satisfaction as a mother and as an intellectual woman in the ways that I have already mentioned. The thought that "stay-at home" women were, in some way, less talented or unfulfilled had no meaning for me, at all.

It is interesting to note, as I look back, just how much deliberation I had given these issues as they were being "tossed about". But I knew how and where I had stood on all of this. In fact, I had just submitted a paper depicting my ideas on the subject of male/female relationships, (including my own relationship), for the course that I was taking in Human Sexuality.

Right now I was well on my way to resuming my professional career, having raised two children with good, solid values, I thought. We were, and had been, undergoing a terrible challenge to our family stability in the guise of a biologically--based, (and apparent) identity problem that my daughter was experiencing. We had marital differences and problems, but in no way was I a "dependent" wife, who was "prostituting herself", "not able to work for even half a day", as I was now being told during the course of therapy. The issues of "giftedness" and the importance for gifted girls to excel, seemed to have become an issue for this therapist. The circumstances of my marital relationship did not seem to coincide with an environment conducive to fostering this, it was thought. So, inaccurate and unfounded views became further "reason" why

I needed to "remain" in treatment.

When I saw that the issues were becoming out of context, I suggesting that the therapy be terminated for awhile, but, I was told that she (the therapist) would tell me "when to stop". I felt really caught up in a bind. If Karen were to experience more of the same problems, (staying out of school), and a risk existed that she could be taken away from me if she didn't meet the school attendance requirements, what chance did we really have if I left "therapy"? The therapist was already seeing me as controlling, and overprotective, (more of the same labels that I had been told had already been submitted to the school system by the other facility). How would Karen or I ever stand a chance if I terminated that therapy? How would my word ever have stood up in a court of law (or in any other arena, for that matter), when I was all alone in my thinking and conclusions and did not yet have the benefit of current research? I could not take that chance. I continued to interact with that therapist, struggling to have her open her mind to my reality that did not need to be "treated." I did not succeed.

Later on, she even blamed our differences of opinion on the fact that the "therapy was too painful" for me. Of course, the "treatment", was painful for me. It was, in essence, right now, destroying me. But I knew that someday we would "get out."

Someday we would get out from under the systems where "authorities" use intimidation and control on those who have to temporarily rely on these systems. But to me, this was a very serious problem--a moral one. I could not believe that prejudicial and accusatory remarks could ever be made in a supposedly therapeutic relationship. The insistence that I "buy" the stated identity (because I had no knowledge to refute it), seemed to me to be incredible. The prolonged accusations were causing further damage to a marriage already weakened. I knew that I could not leave these warped views and systems until my children had gone through them and they were on their own. And yet, it is a good thing that I remained in that relationship, for later on

Karen did develop another malady. It also had no name then. It is now called chronic fatigue syndrome, but then, it was just another example of my "overprotectiveness."

I would manage to get Karen through that new malady because I was able to say, "I am in treatment", (a "treatment" that was rapidly eroding my self esteem). I could not believe the course that my life seemed to be taking.

Senior Year
The Onset of Chronic Fatigue
Another Attack

For awhile, Karen continued to function highly and well in the high school setting, both socially and academically. For many months she had been devoting several hours a week in practice with the State Ballet. When they readied for a performance, practice time tripled, it seemed. The numerous hours spent in practice often resulted in irregular and sometimes, skipped meals--a highly dangerous combination for Karen, but she loved it with a passion! This was in addition to keeping up with studies and other extra-curricular activities. Fortunately, she recognized that this pace would not be a wise schedule to continue, for the up-coming spring season would soon bring the environmental allergies. She was learning about her physical needs through her own experience, learning and anticipating what she had to do to meet them. She cut down on some activities so that she could try some new and different endeavors instead of trying to fit in all of them. She was able to take advantage of the opportunity to work in the summer job which she had been pursuing since February.

She had been on the staff of the school newspaper and had been complimented on her work by many people. Since she loved to write, she quietly and independently investigated working for our local city newspaper during the summer months. This was a marvelous way for her to try out some of her interests and skills as she would soon be thinking about college and perhaps a major in English. She was successful and got the job without anyone's help. The editor was quite impressed by her work, and it was said that Karen was "one of the best student writers that he had known".

From that successful summer, Karen moved right into her senior year without incident. The "buzz" was on about college. Everyone had been talking about it since her junior year, it seemed. We had often spoken about college while she was growing up, thinking it was a "natural extension" for a bright girl, but there had never been any definite serious plan, and she expressed more than once that she didn't even know if she wanted to go to college.

Status or pressure to perform was not a part of my agenda for my children or for me. But we could not escape the almost seemingly obsessive talk about SAT's, entrance exams, and the necessity of making college choices that came from many school personnel and other parents. (And here it was only mid-October.) I thought to myself, how sad to have the qualities of life placed into such a narrow and strong direction with so much emphasis on one goal...college. Although I was mindful of the importance of the many processes and standards in the course of college selection, we did not identify with the many others who were placing an air of crisis upon these seniors. Our value system was always based on who you are as a person, not how many awards you get or how much status you attain...or what college you go to!

Karen, by virtue of her high grades, had remained in the accelerated program throughout high school. As part of this program, she had been assigned to the "College English" course, for students who planned to go to college. For some strange reason, (which I never understood), some of these classes were held in the evening at the teacher's home.

These evening classes did not "sit right" with Karen. She needed early-to-bed school nights. In addition, those particular classes did not seem to lend appreciation to the creative aspects of her writing. Her work naturally received whatever value the teacher wished to place upon it, and often that value was highly under-rated, I thought. I saw that clearly, as I read her papers and the accompanying remarks. But I did not interfere. These were facts of life.

In time, some of the energetic functioning and irregular hours began to catch up with her. I could see her becoming tired once again and even unmotivated, I thought. She was nearing college time, when she would be on her own, and when any of a number of events would happen with which she would have to cope. So, contrary to what some other parents might do, I let the evening classes go and the unseemly remarks on her papers go by without question or intervention with the school.

Although Karen had, up until now, been very mindful of her health and the maintenance of it, she was becoming more tired and it began to show. She was beginning to get colds more often, and they lasted longer. That meant some absences. She cut down on some of her activities and often gave up going to the evening English classes, going to bed early instead. But, in spite of this, she was often out sick, especially during the winter months and, (typical for her), she did not bounce back rapidly. Therefore, when she became ill with the flu, I was not surprised to find that she was not having what some might call a "normal" course of sickness with a "normal" recovery period. (It would have been unusual for her if she did). I thought she was doing fairly well, considering her health history. But this time, it was school officials who "panicked"--not about the state of her health, but about her absence to be ready and involved in the "big rush" for college.

It was "not O.K." for her to be out of school for such a long period of time, especially at this time of year. Seniors needed to be in school..to prepare for SAT exams, to keep up their grades so that they could have the best chance to get into the college of their choice. These were the messages that came to me. This kind of pressure, I thought, usually came from parents needing to uphold a certain image, but such pressure from the school? She had continued to score over-all tremendously high marks in spite of being out. If she wanted to go to college, fine, but there would be no pressure from me for her to go or for her to perform.

November arrived, the time for students to be finalizing college plans and submitting final grades and data for scholarships. And there was Karen, still not recovering from the flu and missing many days of school. She continued to show signs of extreme fatigue and was running a low-grade fever. A two-weeks' absence did not bring about rejuvenation and full recovery. School personnel began to get nervous and demanding.

To satisfy everyone, I took her once more to the allergist. She had long been receiving desensitization shots as well as allergy medications and had been under regular supervision since the early episodes of her symptoms. The allergist didn't think she should be having those lingering symptoms with the treatment that he was providing, but he did not get alarmed. He knew Karen's system. Because of the increasing lack of school attendance, more tests for mononucleosis, thyroid, etc. were done. They proved to be negative. Therefore, she needed to go to school, well or not. We could find no other reason for the extreme fatigue and the failure to "bounce back". The physician could not, at that time, find reason to justify the prolonged absences, but did write a note for her to be excused from gym because of the fatigue. She went back to school, but was in and out sporadically. I was not surprised, seeing this as once again, an indication of Karen's faulty immune system, a fact of life for her. On the whole, she had fared well, I thought.

When she was well, she still showed a marked sense of adventure, motivation and curiosity about life. There was no holding her back in the amount of creativity she exhibited. She had done performance experiments on her gold fish, which she entered in the state science fair (on an individual basis, not connected with any school entries). She made a massive demonstration exhibit about the phenomena of the proclaimed "Doom's Day". As I watched her extreme excitement and high level of motivation at home, I couldn't help but think that some of the depression that she often seemed to experience could be related to a limited expression and appreciation of her creativity within the school system.

I would guess that there was some discussion about her school absences among school faculty. As it was not long before the head guidance counselor telephoned me to say that it would be necessary for Karen to drop out of the College English course. She didn't seem to be able to handle it because of her..."eh, health", were the remarks made to me. I thought, to myself, "what are you people about, anyway? Such inflexibility! Such insensitivity! She was getting excellent grades in spite of the absences. She loved English. It was probably even going to be her college major. And here she was being dropped from the College English course. Is this really what should have been happening to a student, demonstrating excellent writing skills, with a possible future as an English major ahead? Was this supposed to be another "threat" or what? I did not bother to fight. I had fought for long enough and soon it would be Karen's battle to contend with. Nor did I have sufficient information about her health at that time to challenge the decision and negotiate other possibilities. The lack of sufficient diagnosis did not warrant the absences.

I talked to the therapist for the gifted about this. The lack of motivation that I was seeing seemed to be so typical of what I was discovering about unchallenged, unappreciated gifted students. The therapist, still seeming to be convinced that much of the problem was due to the "dependent-wife" syndrome (that she had created for me) maintained that "nobody's that smart". She placed little emphasis on the fact of Karen's persistent low-grade fever, now accompanied by muscle and joint achiness. She insisted that I *had* to send her in if her temperature was not elevated above that which might indicate concern. I could understand that. One is usually able to function with a low a grade fever, but continuously? And for how long? In addition,

Karen's "normal" was usually somewhere in the 97's. When it did rise, that usually meant that something inappropriate was happening. I knew this but I could get no medical explanation and I was again being pressured by school authorities regarding the absences.

It was her third week of absence and she was still not feeling well! Thc guidance counselor and I met to discuss the problem. My attempts to have him understand that these symptoms were usual for Karen seemed to have no impact on him. In his mind, (judging from his comments), Karen was staying out of school for some minor, insignificant reason. He compared her absences to those of severely ill or injured children. By comparison, she did not seem to have reason for the absences. I could understand that, because some of the symptoms, in themselves seemed insignificant. But there should not have been a comparison. It was the whole range of what she had experienced, and for so long, without satisfying explanations, that had been so frustrating and difficult to explain. This was the fifth year. Karen had been to two pediatricians; an ear, nose, throat specialist; an allergist; a family physician and three counselors/therapists, from three different agencies and yet, I could give no reason other than to say that she had a long-term bout with the "flu" and that the "weather" bothered her.

I did my best to recount what I had seen as the problem. I tried so hard to have somebody understand with me what had happened to this bright, beautiful, young woman. But it all seemed so trite when I tried to verbalize it. I realize how it all must have sounded for someone who had not experienced the whole thing with us. But even if there was lack of understanding, I never should have been scoffed at the way that I was. I was told emphatically that children with much worse illnesses got out of sick beds to go to school, especially now that they were in the process of final exams. What was the matter with her? My God, I wish I knew. Nobody was yet talking about chronic fatigue, seasonal affective disorder, multiple chemical allergies and all of those other strange things that affect some individuals more than others. In my innocence and hurt, all I could do was cry and say that "I was in therapy". (I thought that would cover everything.)

When I cried in the therapist's office, all she said was, "don't do that here, do it at school". When I opted for "out" or tried to strengthen my arguments I was accused of being "controlling," and was told, "I'll tell you when to get out". I was locked into therapy which was abusive and degrading to my spirit, yet my daughter remained "hostage" to the system. I would "hang on" until I got her through.

My pain was intense, but my saving grace was the fact that no matter what any one said, I knew that I was doing right by my daughter. Someday, I would realize that, and share it with the world. But for now, I was being laughed at and worse.

My husband was completely exasperated, but continued with his work at the Boys' Club. Karen continued to feel increasingly depressed about her inability to feel better. It was devastating not to be believed and worse still to have the additional punishment of having been dropped from the college English Program.

I began to think about what alternatives there might be. I wondered where else we might take her. And what would that bring? Still more disbelief and harassment? Would anyone else believe me? I did not need or want any further harassment from anyone...guidance counselors, therapists, physicians. I felt so alone.

One alternative (unthinkable to me) was to suggest that she "drop out" of school once she reached sixteen. That seems to have been the course for many people who are creatively gifted. But, what would that have done to her? And would that be right morally? Again I went the medical route. Another visit to the allergist. He again confirmed the condition--multiple allergies. He sent a report to the school, but that did little good. I took her again to the family physician. He did many tests again, "routine" for diabetes, hypothyroidism, mononucleosis and others. And this time I could hear the physician say to the collaborating professional on the telephone, "what's wrong with the mother"?

Nothing tangible could be found from the testing, but the muscle aches and pains had not vanished since the flu, the frequent low- grade fevers, and by now, severe depression, remained. Instead of support, I received ridicule and sarcasm. There was layer upon layer, of disbelief, false accusations and inaccurate diagnoses. Most hurtful of all, I was being blamed for producing, and then aggravating, these nameless conditions.

There had been so many accusations, ridicule and turmoil of my once peaceful home. I had become weaker and weaker in my ability to withstand it all. I felt verbally abused and unsupported while giving my all to help my daughter. I felt wounded with a need for healing. I felt the need for solace and comfort. I would get that later. For then, I only knew that I had to help my daughter.

But there was one thing of which I was certain. If the character and attributes of my personhood, my value system, my morality were seen as things to be corrected and the diagnoses for Karen's ills, by professionals that guided gifted children's potential, then I needed to prove that diagnosis as terribly wrong. I needed to provide a different and more positive outlook on the topic of giftedness. If I did not, other children and families might suffer the same fate. I knew that when this was all over, I would regain my professional status, and that I could-- and would-- do something to prevent others from suffering in the same way. I knew that I needed to distance myself and to retreat in order to heal. I would do so once I had "seen" Karen through.

It wasn't until several years later that my knowledge and research about Karen began to match what I saw appearing in scientific journals. Later still, I noted an article in our state newspaper decrying the failure of pediatricians nationwide to recognize chronic fatigue as a reality in children. And it wasn't until much later, through contacts with a researcher at Yale University who was studying neurobiological disorders in children and adolescents and with researchers in the field of nutrition, that I confirmed some of my earliest observations and assessments and also that dietary components really could positively resolve some of these conditions.

College, a Choice....Whose Choice?
A Turning Point for Karen

Karen, did her best to follow her diet, get appropriate rest and put all of her existing energies and efforts into getting herself through the school year. She developed her own sense now of when she was getting the feelings of uneasiness, accompanied by sweating and faintness. She found that mid--morning and mid afternoon snacks could bring some relief. She began to listen to the needs of her body. The long-term, chronic fatigue and residual knee pains mandated that she give up what had been her life-long love of dance. But her developing maturity handled that now and her intelligence was devoted to planning a course of action to get well, finish her senior year and function in accordance with her limitations and assets.

Unknown to me, Karen had been exploring colleges. One day, she presented me with the name of a well-known, out of state all-female college. She had researched it and thought that she would like what it had to offer. I was surprised, but really happy at the one she had selected and agreed to help her with whatever she needed to get admitted. Grades were no problem. In spite of everything, she had done well on her SAT's, and had maintained all high-ranking scores. I knew she was intelligent and studious, but this, to me, was astonishing...that she had maintained such high grade levels in spite of the many rebuffs and obstacles that she had faced. She really cherished life and learning.

Fortunately, the college of her choice recognized her gifts. Even though she had so many obstacles through the years, she had still managed to accumulate an impressive array of grades and community activities. In addition to her many years of dance and little league, she had been in girl scouts, had won journalistic and science awards, had volunteered with the children's program in the local library, and had been a student campaign worker for a political candidate. She had even delivered newspapers and at one time sold cosmetics door to door. There was nothing wrong with her motivation. All she had to do was learn to balance an intensity of spirit with the limits of her physical condition.

Not only did she receive a letter of acceptance from the college of her choice, but she was also chosen as a named scholar. She ranked in the top 5% of applicants for that college and would be awarded, while there, the use of a highly popular, expensive computer, which would be given to her upon graduation. All this was in addition to receiving named scholarship funds. Absolutely remarkable! I was extremely happy for her. She seemed to be well on her way and constantly gaining in maturity and understanding.

Karen herself was beginning to get disgusted with the school systems. She did not need the reprimands that came from school officials regarding lack of attendance because of her health problems, (which she knew were real). She was also aware of the consequences that her problems were having upon our family. Years had now passed since we had functioned as a united, happy family. At a time of adolescent growth, when most families are in need of support, ours was instead being torn apart by the very "systems" that were supposed to help. We were dealing with a multitude of symptoms and a conglomerate of professionals, lacking in enough knowledge, controlling in nature and much too quick to make us own the absoluteness of their labels.

In spite of the academic scholarship and other honors that had been awarded to Karen, she made the decision not to attend college. She wanted to take stock of herself, to renew her life and to ponder future goals for herself. Not a bad thing for a teenager to do! (It was not for many years later, after she had assimilated and processed her several years of experience, that I learned just how much the restricted use of her creativity had to do with her constant low energy levels, and how much the deeply imbedded traits of honesty, forthrightness and justice, (traits which are often shunned and ridiculed in our society, but seemingly inherent characteristics of the creatively gifted,) had influenced her growing-up process.

It is said that if one is not growing, one is dying. Spirit in creatively gifted children is a *very* strong force. It has to be nurtured. I do not believe that school officials know or understand this. They often seem so intent on *driving* the students in one direction, not realizing that sometimes "bright" students have needs which vary from the "others"--and that they are not desires or options, they are *needs*. They also do not seem to realize that acceleration of programs is not necessarily synonymous with the promotion of creativity. Nor do they seem to understand the drastic need to provide opportunities for growth and construction...necessary to foster potential. In other words, "more" is not necessarily better". Accelerated programs, I would guess, benefit the more verbally gifted versus the highly creative who need to invent structure rather than follow it. As a parent, I do not expect school systems or any other system, for that matter, to meet the needs of every single individual. It is sad enough that they have had to assume the long-ago relinquished parental roles of feeding breakfast, teaching sexuality, giving driving lessons and keeping guns out of school. And, in fairness to the schools, I have only gained all of this knowledge through lived experience. But the one expectation that I do have is that "systems" treat people's individual differences as "differences" and not destroy their potentialities with labels and assumptions that differences are negative characteristics to be "corrected".

It is also to be noted that practices and beliefs varied among teachers. For the very system that brought Karen so much difficulty with regard for learning was the same system that had once called her "gifted" and spurred her to high levels of accomplishments.

Finally,
An Integrated Accurate Assessment

My daughter had survived the trauma, but she had a lot of life yet to be lived. We really needed some medical confirmation of all of this and to accurately integrate all that had happened! But I dreaded the thought of telling the story, even one more time; I dreaded the possibility of being scorned, ridiculed and negatively labeled even one more time, but I loved my daughter enough to try, even one more time, to get this whole situation integrated for her. I needed to find something other than the word "allergy" to account for her problems. Since nobody in our state, (that I had found), had been able to provide that. I made an appointment for her to be seen at a well-known Boston hospital.

We were assigned to a young female physician. It was very difficult to tell that long story again. And each time I told it, the terrible complications, adverse reactions and accusations that had been heaped upon us along the way made it lengthier and even more painful for me. My emotions were raw. The young physician saw that and listened intently. Her very first comment was "don't feel so guilty". I took a moment, gratefully to believe, that there was really someone who could look beyond guilt and blame.

As she spoke, I realized and was able to verbalize the enormous amount of guilt that I had been forced to absorb throughout this ordeal. I had not felt "guilt" when the problem had first occurred. I had done nothing to feel "guilty" about. I had done things earlier in my life that I would do differently another time around, but I couldn't feel "guilty" about things that were done to the best of my ability at the time. In fact, I remember telling the young staffworker from DCYF, at the initial stages of the problem, how this was probably another "blessing in disguise". I had always looked at life that way. Things that were difficult had always turned out to have some positive aspect-- not easily, but they had. I did not for one minute think of myself as the cause. True, I had smoked during my pregnancy with Karen. I was sorry that I had done so, but I never felt "guilty" about it. There were also times in my life that I had felt sadness at others' response to some of my actions during the years, (as a person very sensitive to others feelings), but I did not feel "guilt". I was sorry that they did not understand, but, I did not feel "guilt". I had always acted according to the dictates of my conscience. I had done a *lot* of things that I was proud of, too, and it was with that sense of healthy acceptance that I interacted with others.

To me it is a wise person that knows and can admit their mistakes and weaknesses. I had reported my shortcomings to the therapists from the very beginning, never expecting those issues to become distorted. My remarks were amplified and misused. I was not "allowed" the humanness of experiencing the paths of mistakes-- in essence--the paths of learning. Throughout our journey, there was much misplaced and projected anger, guilt, blame and insecurities, yet these were not my perspectives. Someone else's thoughts and feelings had been forced upon me and I had been forced to absorb them. I even remember having held my hands to my ears, trying to block out some of the accusations. In spite of it all, I had carried on, but it would take years for me to dissolve that feeling.

In contrast, it was so wonderful to hear the young physician in Boston saying so simply, "don't feel guilty".

She recognized immediately that Karen was very bright and that she had been very, very tired. She ordered many tests, finding elevated cholesterol and triglyccride levels. (That was before the discussion of cholesterol had become such a big issue) and one more thing indicating that her energy level was being affected with hormonal, metabolic implications. The sweating, trembling, and heart-racing sensations were traced to a fluctuating low-blood sugar level. The lactose intolerance was verified for just that, an intolerance, not a "nervous mother." The end result was that she needed to follow basically the same diet that had been originally prescribed years ago (for "allergies"), and to get plenty of rest. Above all, we were not to feel "guilty". Karen, she thought, was a very special young woman, --and it showed.

Karen did not have major problems, but a conglomerate of many problems that, during puberty, took more energy than what she had available. No wonder I had seen such vast improvement and "healing" when Karen followed the diet that had been prescribed for "allergies" when she was 6 years old. When we reviewed the diet which the physician from Boston prescribed for the high cholesterol and triglyceride levels, it coincided almost exactly with the earlier one. Later research confirmed for me that she had benefited from those extra amino acids and vitamin and mineral supplements, probably having needed them from the day she was born. Later information also confirmed the importance of some of those basic dietary substances in the making of chemical neurotransmitters known as serotonin and others..for which our needs vary. There simply just was not enough available energy for what she needed; especially so, during her teen years, when an even greater amount was required for growth; for processing her identity issues; for combating the effects of environmental allergies; for making hormones and for taking part in all of the activities in which teens wish to participate. She was experiencing demands from a variety of sources, the demands of a body requiring and seeking what it needed and did not have. With this accumulated information, we can now confirm what the real issues were:

1. Getting the appropriate kinds and amounts of foods/nutrients that she needed (She had discontinued most of the nutritional vitamin/mineral supplements at a time when they were of the

utmost importance for her.)

2. Experiencing the additional strains of hormonal changes while experiencing deficiencies of the above
3. Experiencing the stresses of being a highly impressionable and sensitive teenager, in both the physical and social contexts, and brought to crisis proportions by a sensory overloading during the camp experience.
4. Experiencing a process of identity formation in a social environment (school) which did not fully recognize her needs and abilities and, therefore not adequately respect them in fostering her growth.
5. Being in a society which still does not know/ appreciate the enormous difficulties that are experienced by environmentally sensitive people, people with "unseen" biochemical problems, people who are potentially creatively gifted, and others who are just different from the "norm" in any way that is not visible - in short, as the therapist so accurately named, but failed to pursue, - the "hidden disabilities".

Evaluating the Devastation

The school year was over now. Karen had not been taken away from me. That was a most important accomplishment. There was no further need now for the "counseling" sessions as we had known them. What we needed now was help for the damage that had been wrought upon our family. I left the therapist, with whom I had shared so much. We left with un-resolved issues and parted in a very cold and distant manner.

I was left with a feeling of total isolation for having fought this battle mostly alone. The "therapist" had made many unfair accusations against me, (only some of which are mentioned here), my role in the marriage, my lack of professional status, my "need to control". Little by little, she had torn down everything that I valued about me, including my ability to remain committed in a relationship--the "therapeutic" one.

And so my relationship with that therapist ended with each of us on seemingly different planes. I left in a state of total frustration, anger, depletion of resources and with deep emotional scars. Later requests, (by me), that we meet to discuss the process and where errors had been made, were not granted. After what I thought was a sufficient period, I took the alternative of expressing my anger and displeasure via written correspondence. (I was aggressive.) I also sent her letters, research articles, and news clippings as I uncovered pertinent information. As much as I would have liked to have resolved our conflicts, I had some consolation in the knowledge that I had sent her new, correct and current information about the medical components of Karen's condition, previously mentioned. That has helped me to put closure on that part of my life.

The early differences that my husband and I had experienced

had grown much larger. Intellectually, I knew that my husband handled problems in a much different way than I did. His emotional make-up was and is different from mine. I wish that such was not the case, but it is. Everybody acts or reacts with life's circumstances in various ways. The tragedy is that during therapy, (contrary to what I had always done), I was encouraged to put negativity and distance upon relationships in which others and I had differences, instead of helping to diminish those differences through growth and understanding; I was encouraged to personalize those differences in behaviors with labels, rather than to address and try to resolve the underlying issues involved; and to use energy in reacting to differences instead of pro-actively getting on with my life. In other words, I was taught to emotionalize, rather than rationalize. (I did not know, nor could I verbalize at that time, what was happening, but I did know that her way of thinking and relating was very different than mine.) It was only after much sorting out was I able to arrive at those insights. As a result, divisiveness began to happen in so many of my relationships, as I carried out her way of thinking, to determine what had transpired. I really felt that I had been brainwashed. I felt myself and my value system beginning to crumble from the prolonged attacks and faulty attitudes towards me. I felt left with numerous derogatory labels on me and even with a false identity, in my own community, where I had previously enjoyed pride and respect. What had happened was not right. I felt alone and unsettled in relationships.

I did not know anyone who had gone through this particular experience. There was no one to whom I could relate about this particular experience. The names of agoraphobia and chronic fatigue syndrome still had not yet appeared, and I did not know of anyone who had to undergo what I had in the name of "treatment". It was bad enough that my daughter's problem had occurred, but worse still, were the traumatic results of the prolonged "therapies" intended to make me over. I could not leave circumstances and beliefs as they were. I felt as though I had to make a choice between two evils:

1. to leave the labels, diagnosis and false impressions of my daughter and I as they were--or
2. to take actions that would disprove the diagnoses, restore my good name and character and seek the truths that would rightfully explain the problems. This route would also clear the way for others to make it through the same journey, if need be, in an easier way. I could not stay at home and in my community with the distortion that I felt within me.

There were several things that I felt I needed to do:

A. leave my husband in order to have the distance and solitude to work this all out (erasing the labels of dependent, etc., while, establishing a peaceful way to co-exist in spite of the differences between us.

B. leave my daughter (and son, as well) erasing any last vestige of the term "overprotective mother" and allowing them the ability to exercise whatever lifestyle they chose (in view of the differ ences between my husband and I) and

C. regain my professional status so that when the truths of newer research came, I would be ready to "run" with them.

D. regain my professional status in order to "prove" that "stay-at-home" Moms often have capabilities that are often equal to those of other professionals "out there."

The biggest heartbreak for me was the fact that I felt that I had to "prove myself" and to "right" the situation. I had never in my lifetime felt that drive so strongly, but I could not forget that, because of some foolish rules, someone might have had the power to take my daughter away because we lacked sufficient knowledge, (at the time), to keep her--even if a court battle had ensued. I knew that I could not continue without some sort of order being made out of the chaos that we had experienced. I knew that I needed to gain the knowledge and to pass it on. I had been forced to take actions which, ordinarily were against my better judgment, but carrying them out decreased the chances that my daughter could be taken from me. I had been forced to absorb words that did not apply to me. I felt separation from the God that I once knew.

The separation between my God and I was caused by words--mound of words and tones that got in the way. I was "buried" under a heap of words. Time and time throughout life, I had been buried under words that said who I was or who I should be. But time and time again, I had freed myself from those words and each time that I did, I was able to verbalize more accurately the truth of me. It was these self-discovered truths that had provided me with knowledge to help others in their growth. And now, once more, words had been put upon me that were untrue for me, that did not describe my truth and yet were the views of present "authorities". And this time, I knew I needed to eliminate those words that had been heaped upon me to discover new truths and new words. I had to wait for truth to happen. There was yet another spiritual path to follow.

One, by one, I walked through the inaccurate words that had been placed on me. I believed that when I had finished this task, I would once again be in harmony with myself and in union with my God and the freedom that goes with that. In other words, united in truths.

Beginning the Journey Back

I became employed as a "vocational specialist", dealing with the tasks of job-finding, job development and counseling for the rapidly growing Southeast Asian refugee population. This job helped to begin to eradicate the therapist's simplistic remark that I "wasn't a professional yet" and that I "wouldn't be able to work for even half a day". I worked long, irregular hours. I could be found in factories and in homes, transporting and teaching. In spite of the hours, hard work and frequent frustrations of not being able to find enough jobs for these people, there was a satisfaction that cannot be found in more "traditional" jobs. I enjoyed the many aspects of that population and learned much about the culture of Cambodian, Laos, and Hmong people. I reconfirmed my work ethic and my comradery with my fellowman.

Meanwhile, I was still enrolled in the college program and had completed several courses toward my degree in Agency Counseling. I was now nearing the end of the six-year time limit allotted by the school for completion of the program. I had two courses left, one of which was the second Practicum in Counseling. This course consisted of the actual counseling of an individual and/or family. During this time a professor guided, mentored and critiqued us in the development and/or refinement of our skills. I needed to be take this course to complete my program but how could I take it now? I felt it would not be fair for me to enter into a counseling relationship until I had renewed myself. And I knew that renewal needed to take place in solitude, away from my family. I decided to take a one or two semester leave in order to do this. I would complete this course later.

I did not know how my husband would react to my leaving, but I knew that I could no longer stay with the anger and sense of isolation that I felt. His pride and joy had been his family and his children. And here I was breaking up our home, or had it already been broken up? I believed and hoped that my two children could pick up where I left off. They were close to adulthood, and I hoped that they were strong enough to carry on successfully. By now, I had given them all that I could in the way of "upbringing". And so we each had to grow one more time in our own separate ways.

I moved to a neighboring state and to a community in which I had once worked. I felt that there, at least, I knew where emergency services could be obtained. There were also one or two people in a nearby city whom I had not seen in a long time, but who I knew would be there for me if I needed them in a crisis. But I did not even tell them all that had transpired. It was just too "big" and complex.

I was busy making that move when my colleagues in the Master's program were anguishingly preparing for comprehensive exams. It was odd to see the emphasis that was being placed upon these exams by the others in the program. My life situation seemed to be so different from theirs at this particular time. Week after week, they met in study groups to prepare for that big day. I, too was concerned about it. With all my heart, I wanted to pass those comprehensives. I had put so much into obtaining that degree. And I knew that I would be a good counselor, but I could not get involved with them in the long-term studying sessions. I had to take care of my priorities, such as getting a job in my new location and paying the rent.

I remember spending two Friday evenings on the floor of my new-found, but yet-unfurnished apartment, studying. I had no trouble writing about my beliefs and philosophies of counseling. On those two nights, I had meditated on my concept of man in preparation for my exam. I was easily able to integrate my beliefs with the counseling philosophies that I had chosen. I thought about my ideas on the value of man. On the day of the comprehensive exams, I put them into words. I passed with flying colors!

College "Rules"
(Does Learning Only Occur in Classrooms?)

I did not register for any course during that next semester nor did I make any communication with the college. It was not wise to take the "Practicum II" under the circumstances which I have described. I needed time to sort out my feelings and the situation that I was in. I needed to justify the outcome of that situation. If the answers were not available for us, then, someday they would be. When they were, I wanted to be ready and updated as a professional so that I could teach, educate and help to eradicate the erroneous widespread opinions, which I had encountered and which, I find, are still prevalent. But, I also needed time to heal and to try to regain order and, hopefully harmony to my life.

I left behind all of that I loved... my son, my home, my family, my friends, my community. I had left behind most of my possessions except for my most prized ones...my guitar and my bicycle which were to provide me with much-needed diversion and comfort.

I kept in touch with only a few close friends. Some of the previous contacts that I had changed, as had, my circumstances. Some could not understand and were unable to "hang in there" with me. Besides, they had their own burdens of life to contend with.

I was fortunate to find a small inexpensive, one room efficiency unit. It was part of a multi-apartment complex, set in a beautifully landscaped wooded area. Provisions for security were ample and there was an outdoor swimming pool. I did not mind that I was in one room or in a basement apartment or next to the laundry room. I furnished it to be cozy and filled with the warmth of small objects that were meaningful to me.

But my heart ached each time that I thought of my bright, caring and sensitive son. He had only just completed his junior year of high school. I was not at home for his final year. In my wildest dreams, I never would have visualized that this could have happened. I would see Karen often, and she knew exactly what we had experienced together, but my son did not. I met with him as often as we could. The meetings were awkward, he said little and asked no questions. All he knew for sure was a chaotic mess was growing in our household. I cannot describe the pain and anguish that I experienced knowing that I had left my son. There were no words to sufficiently explain what had happened. There was no explanation. I had just left.

One of the first things that I did was to register at the local "Y". I knew that I had to keep up my routine exercise program and did. Although I could not feel the rhythm of my soul, I made the movements of swimming my laps at least three times a week. I usually arose early and attended the 6 A.M. swim before work. I met some fine people there who did not know my circumstances, but who befriended me.

It was difficult meeting people for the first time. At this time I had not yet re-established my nursing role and really could not name myself as a wife and mother (unless I added ...one who had abandoned that identity). In spite of the many derogatory remarks that had been levied against me in therapy, I knew that I had many professional skills, abilities and accomplishments to my name. I was lucky to have maintained contact with a few professionals who knew that and supported me with recommendations regardless of my circumstances. With that support, I set out immediately to find a job. I qualified for quite a few.

Very soon I was accepted for a position as a Quality Assessment Specialist in an agency that serviced the elderly. There were many employees there who readily accepted me without regard to what was going on in my personal life. In spite of my emotional state, my nursing and counseling skills remained strong and served me well there. Part of my job consisted of carrying out mediation between a staff of about 30 case managers and providers of service to the elderly. I found much warmth and support among the members of that agency, but as the months went by and plans were being discussed for the Christmas season, I cried often for the lifestyle and family as I had known it.

During that time, I received a letter from the administration of the college. It confirmed that I had not registered and was not scheduled to take a course. It informed me that established policy dictated that I complete the degree program in six years. My status at the college would have to be terminated. I thought to myself and out loud (to at least one other person), I have been out there learning more than what could be taught in any program, but it doesn't constitute what I need to qualify for the required certificate of completion! Fortunately, I was allowed to appeal.

With the help of the appropriate college administrator, I was able to negotiate an agreement whereby I could complete my degree requirements. I would have to enroll for the next semester. That would be the final extension! It was with many mixed emotions that I registered to take that course of Practicum II in Counseling. How would I conduct myself? Had I resolved my issues sufficiently to participate in a practicum such as this? To be there would be to find out. Not to be there would be to sacrifice all that I had done toward completion of my degree. I was most happy and indeed fortunate to discover that the professor teaching that particular course was the one who had also been my mentor for the prerequisite course of Practicum I. Due to the nature of those practicum experiences and to her tenure in that program, she had much insight into the abilities and sensitivities of her students. Not only that, she had a strong professional relationship with, Robert Carkhuff, "one of the one hundred most-cited social scientist of our time", (grouped "with historical figures as Freud, Marx, and Dewey")[4] and the author of *Toward Actualizing Human Potential*. She knew well of that which she taught. She lent that aspect of humanity and knowledge to us in a most important way.

With some trepidation, I attended the first session. We were instructed about the nature of the course and its requirements in terms of knowledge, practice and general demonstration of skills and abilities. I talked to the professor privately and briefly about what had happened in my life since Practicum I and my concern whether I would be able, at this time, to give my counselees the most that I was capable of giving. I did not know if that could happen, but I also needed to complete the assignment this semester. Certainly I would try my best toward that accomplishment. We agreed that I would not counsel an "assigned" family at this point, but she said to me, you're going to have to come up with something pretty creative. I did! And we did! I counseled Karen. We called my Practicum "Counseling the Creatively Gifted".

It was not easy. It was excruciatingly painful, yet rewarding and fulfilling at the same time. I was able to keep my own feelings about the situation separate from the counseling process. They did not seem to interfere with the knowledge and skills that I had assimilated and had already used to help Karen through the after-effects of her experiences. I did the course requirement of self-critique from our recorded sessions. They were critiqued again by the professor, who had acted as mentor throughout this and who, ironically, and fortunate for us, held the D.Ed. Degree. In addition, my beautiful daughter, (who still always wanted to be involved and to help), critiqued them on an on-going basis from the standpoint of "counselee". (I believe I was the only one in the program to include feedback of the counseling process from the "counselee".)

Karen benefited from our sessions. They helped her to work out some goals and plans, one of which was based around her decision of not going to college. She wanted some time away from a school environment to sort out her life and also to accumulate experience in another aspect of life, a most important one, that of being responsible for one's self. I certainly could support her in that!

Growth "Happens"

Our family members continued on, each in his/her own way. My son was now a senior in high school and doing well. He had secured a job as a salesman and general helper in a large department store and worked long hours after school. His early love of music seemed to help sustain him throughout this most difficult period. He purchased a guitar and was taking private lessons. He was thus adding to his performing repertoire, while he continued to perform as a trombonist with the school bands. He began a course of independent living, further developing the musical abilities that he had demonstrated with such proficiency so early on. He applied himself and his discipline to that music. He became the band's drum major that year (his senior one) and at the end of it received the first John Philip Souza award that the school has ever given. I was proud of the way that he was maturing.

Each member of the family was proceeding onward, but we were no longer the "family" that we had known. Fourth of July came and went. I had not been able to bring myself to attend the large family gathering which I had initiated years ago. I was alone and seemingly without relationship. When Christmas morning arrived, I came in from out of state to be with my children as they opened their gifts. My husband had made an early morning trek to the country to get Grammy so that she also could be with us. Later that day, I spent time with my brother, my sister and her family, and my mother. I was there, in body and in reason, trying to bring back the pieces of what once was, but I am told that my "spirit" was missing from those occasions that had once been so meaningful and happy for me.

My 50th birthday came and went. The joviality and celebration that I had long anticipated for when I became "over the hill" did not happen. I could not even honor my family's request to go out to dinner. I was not in any way, in a celebrating mood. But, true to character, my daughter would not let me forget this special occasion. She appeared at my apartment with a gift and a balloon on which she had drawn a picture of me going "over the hill" on my bicycle.

The following month, I fulfilled the requirements for a Master's Degree in Agency Counseling. This, too was nothing like what I had planned. I had so wanted my husband's family to share in this special occasion. I would have been the first of that particular family of first and second generation Italians to obtain that degree... and I was a woman. They had respected and supported me in this endeavor and I, in turn, wished to share that honor with them. But, such was not to be, separated as I was from them. My faithful daughter attended, bringing along her father, at my behest. I wanted him to experience that part of a world in which I was involved and which, I believed, that someday my daughter would also be involved.

I missed my son's presence at the graduation. I had not had much contact with him. He had moved to the ocean area of the state, close to his college and near the groups of musicians with whom he worked. Life was nothing like what I had planned. Years earlier, we had joked about the fact that Mom would be graduating in the year between the two kids' graduation. But, since the split, we had not joked. My son and I had not yet discussed the troubles of that period and the impact that they had upon our family. I only hoped that he realized that I needed to be on this journey at this particular time.

Karen and I saw one another fairly often. She was working at a furniture store and seemed to like being self-sufficient. That year away from school had given her time to ponder her life and to make some focused decisions. She had gained valuable life experience and earned some money which she would put toward college expenses. Yes, she would go on to college.

She still had the liabilities that I have described herein;

susceptible to environmental conditions; needing to remain on a strict diet and get a greater amount of rest than what her peers required. But now she was older and had learned how to cope with that. She was no longer troubled by feelings of "differentness". She had a comfortable sense of her unique identity and related very well to people, both children and adults. Moreover, with her own high level of intelligence and maturity, she developed a regime which enabled her to carry on a program which integrated work, fun and relaxation.

Karen was aware that a full time course of study and dormitory living at the college might not be the most advantageous for her, in view of her physical status. Since the policy in effect at that time dictated that students live on campus and attend classes full time, she went before the college board to present her case. She requested a waiver of the policy and was successful. In addition, she was assured that she would receive the full benefits of her previously-earned awards even though she would be attending classes on a part-time basis and commuting. We could not have been happier.

She and I were very close. Although she had remained at home with my husband, we saw one another often. She visited me at the apartment frequently, often staying overnight and I still went fairly often to the "homestead", continuing to carry the responsibilities of budgeting and household management. The house itself had changed though. It now seemed like a shell. It no longer felt like the home into which I had put so much love and effort. It seemed that my home of love had been broken, just as my spirit had been broken.

I conversed with my husband when important issues needed to be discussed, but I was still full of the hurt and pain of the relentless, wrongful accusations that had been made against me and my marital role. I still held a lot of anger and resentment. I did whatever needed to be done at home and then always returned to the "aloneness", the quiet serenity of the haven of my apartment.

I steadily re-entered the spheres of my professional life in which I had once flourished. I regained my academic identity and represented the 1965 graduating class at the sesquicentennial celebration of the university from which I had received my Bachelor of Science Degree. On that weekend, I marched in the parade of classes and took part in other celebrations which were also attended by, then president, George Bush. Once again, here were special occasions and no one to share them with but my daughter. I could not enjoy them with my former enthusiasm. But I was, each moment, healing and moving nearer to the time when I would tell this story as a professional with my own credentials and my own truth.

A Legacy

Karen and I continued to develop in our own roles, each in our own right. As she steadily resolved her issues, she gained in proportionate strength and maturity that would be steadfast in her for life. My role in bringing her to that level of maturation was nearly completed. She was ever mindful of the journey that we had traveled together, but also alone. She was proud of what I was accomplishing as a person and as a professional, and she always let me know that. She often expressed her gratitude and it was heartfelt. She left me with a special legacy that says it all.

One evening, long after she had begun college, we both happened to be at home at the same time. She asked me to go into her bedroom where she had a tape all set to play. She sat on the bed and motioned for me to sit beside her and said, "Mom, you have to hear this." The two of us sat very quietly listening to a song that I had never heard before. It was a song which was to become very popular. It epitomizes all that transpired between us and goes like this:

"It must have been cold there in my shadow.
to never have sunlight on your face
You were content to let me shine, that's your way
You always walked a step behind
So, I was the one with all the glory
While you were the one with all the strength
A beautiful face without a name - for so long
A beautiful smile to hide the pain.
Did you ever know that You're my hero?
And everything I would like to be
I can fly higher than an eagle

Cause you are the "Wind beneath my Wings"

It might have appeared to go unnoticed
But I've got it all here in my heart.
I want you to know the truth - of course I know it
I would be nothing without you.
Did you ever know that you're my hero
And everything I wish I could be?
I could fly higher than an eagle
For you are the "Wind beneath my Wings".

Did I ever tell you, you're my hero?
You're everything, everything, I wish I could be.
Oh, and I can fly higher than an eagle
Cause you are the "Wind beneath my Wings"
...the "Wind beneath my Wings"...
Fly, Fly, Fly away
You let me fly so high
So high against the sky
So high I almost touch the sky
Thank God, Thank God, Thank God for you
The "Wind beneath my Wings".

"The Wind Beneath My Wings"
(Jeff Silbar, Larry Henley)

Mom and Karen "On the Move"

Karen successfully completed one semester of half-time study and then went on to complete a semester of full-time study. Her creativity, maturity and love for learning was being re-ignited, and in turn, it was appreciated and nurtured by the professors at college. She again earned high grades and praise from teachers about the quality of her work. She was recognized with the awards of named scholarships, which were very helpful in supplementing loans that she was accruing. This had been the college of her choice and apparently was an excellent one for her. Her work and enthusiasm seemed to parallel her earlier years while in Project Gifted.

She had learned to schedule her work and even resumed dancing, taking advanced ballet lessons at a local studio. Her spirit was returning. She was ready to try the challenge of attending the last two years full-time on the college campus. She had the routine down pat. She was doing well and she would succeed!

The policy dictated that single dorm rooms be given to those who could substantiate, through medical or other reasons, why they required a private room. She went to the allergist who had treated her for so many years, to request such a recommendation. I was astounded at the letter that the physician had written. It read:

"Karen has been a patient of mine since 1974 for multiple allergies. She has been on injection therapy since 1982. While in high school she demonstrated the typical seasonal "tension-fatigue" syndrome in the spring and fall. She has improved somewhat but still experiences extreme fatigue throughout the year. In 1985 we ordered extensive blood tests including CBC, Immunoglobulins, Thyroid, Mono, SMA Profile and a Urinalysis. No serious disorders were found but some of her symptoms remain. It would be our opinion that because of her allergies and her required sleep habits that she should be given serious consideration for a single room at school. In long-hand was added, "Patient needs a greater amount of sleep related to her extreme fatigue...needs extreme daily rest and uninterrupted sleep."(!)

How I valued that information. How I wanted to post it throughout the high school system and fly it around the world to all of those people who doubted me, who laughed at me, who scoffed at my knowledge, who had made a total mockery of my parenting. But I would not emote in that way. I remembered the vow that I had made, that once I got through the emotion of that ordeal, I would educate about these topics with the high degree of professionalism that I once knew.

Meanwhile, I found that the job I was in was no longer suitable for me. It had provided sufficient income and a warm work atmosphere. I was grateful for that and the support, companionship and tolerance of my low spirit while I was there. But now I needed to leave. My role, as it was defined, limited me in being able to "regain" the professional nursing status that I sought. The organizational structure of that agency did not allow for the particular advancement that I needed to secure. I began to look for another job where that might happen.

After only a few interviews, I was hired by a large private rehabilitation agency. It serviced people who were unemployed due to job-related injuries. In this position, my nursing status was recognized and for the first time, since my return to the work world, I was able to work under the title of an R.N. My combined education and experience in the nursing field, plus my more recent experience and education in the vocational field stood me well. I had also passed a State Civil Service exam for Rehabilitation Counselor, which added to my list of professional accomplishments. Now, I had accrued job titles of Vocational Specialist and Rehabilitation Specialist, with an R.N status.

The position for which I was hired placed me in working relationships with physicians, attorneys, rehabilitation facilities, the

Workmen's Compensation Board, the injured workers' insurers and their employers. The tasks required making nursing assessments, rehabilitation, mediation and negotiation with professionals from many different agencies, job development, and the use of counseling skills to foster re-employment. This was a perfect setting in which I could use my integrated skills.

I liked the aspect of independent scheduling and the challenges of the job. But, after only a few months in this job, I again felt limited in the scope of how I could apply my knowledge and experience. I could not focus on healing the whole person, but only on those aspects which related to the sustained injury and for what the insurance company would pay. I was supposed to see only that "part" of the person-- apply a "bandaid" and get the person back to work. This was very difficult for me. I had trouble "ignoring" other aspect of the clients. It was obvious to me many times, that the worker probably sustained an injury because of other factors. But, in this job, I was not allowed to act on what I viewed as total needs. Not to practice in harmony with my knowledge, education and belief system was not appropriate for me. I realize that businesses need to be profitable, but I could not compromise what I felt was fair for a client in terms of what I could provide toward a more permanent health status. Circumstances evolved which necessitated that I leave the agency, before I got fired for my lack of compliance with the required policies and procedures. And so, with two children still in college and other bills to be paid, I found it necessary to terminate that employment. Others would call me a fool, but I had to practice according to the dictates of my beliefs.

It is fortunate that during this time, both children earned academic scholarships as well as meeting some other financial aid requirements. Both also worked at various jobs to pay for a good portion of their college education. It would have been much easier for me to have remained in the same place of employment for a much longer while, but I knew I had to look again.

My resume now reflected numerous and varied professional

roles; program planner, coordinator, supervisor, administrator, innovator of new programs, (with most of those jobs having taken place with an extremely diversified population). The groups, with which I had worked, came under the titles of "elderly, chronically ill, unemployed, Southeast Asians, American Indians, people who were 'disadvantaged' socially, culturally, emotionally, economically; those with substance abuse problems, mentally and physically challenged, single mothers, youth groups, parish groups, those with diverse sexual orientations, those with "reading difficulties" and many others. The issue was not that of finding a job which I could do, but finding a person or agency who would allow me to work with my unique combination of experience, education and values, and reimburse me accordingly. Up to this time, I had worked in human service agencies for "below-salary" because I did not hold the "correct" piece of paper for many specific roles. But, in the meantime, I was advancing professionally and beginning to heal.

Return to Home

I had so many thoughts and memories of the home and family that I had created. I had lived away for almost three years now. Since that time, Karen had completed three years of college, completing her final two years on campus. My son was still living in the vicinity of the college, which he was still attending. The only one with permanent residency at home now was my husband. It was as good a time as any to try to renew the life and relationships that I had known.

I returned home.

I returned to my routine lap swims at the pool where I had gone regularly for some twenty years prior to my departure. I resumed contact with the other women swimmers. They were glad to see me. They asked few questions about my long absence, although they knew that something strange and strong must have prompted my departure. But they did not ask what. I was able to resume those relationships and was treated me with the same admiration and respect that I had always received from them.

I tried to put myself back in time and place to where I had been prior to leaving home. As I had gone through the healing process, I had discovered something. Whereas my spirit had usually led me, it now seemed to be the other way around. Once I performed an activity, the spirit for it returned. And so it was, that I began to, once again, become involved and little, by little, experienced the fun things of life.

I read an announcement in our local newspaper, that a Senior Olympics would soon be held at a local university field. I had never before attended that event, but my curiosity and eagerness to get on with my life, prompted me to go and watch. I stayed for only a few events, observing the female participants walking and racing, and said to myself, "I'd like to do that. Maybe next year, I will". I returned home, eager to evaluate my running skills. I set out to run the length of my back yard only to find that my legs would not run. I couldn't believe it! I swam and bicycled regularly and had been doing the U.S Air Force daily exercises for some 15 years now. How come my legs would not "run"? I guessed that the "run muscles" had not been used in a long time. Not one to succumb to defeat or regression unnecessarily, I thought to myself, "come next spring, I'll practice running". In the meantime, I would make it a practice of running short distances. So throughout the year, whenever I thought of it, I could be found running from car to door and from office to car. I was definitely beginning to show signs of healing.

Life Goes On

We were all beginning to see and experience some positive things happening in our lives. My husband, now nearing retirement, was beginning to receive recognition, locally and nationally, for his many years of service to the community. There was even a State Day named in his honor.

My son was diligently pursuing his career in music. He advanced in theoretical knowledge of music through the college courses, while learning the intricacies of managing sound systems from an apprenticeship, which he had arranged himself. He combined his many talents in developing his own self-styled career. He began to play both trombone and guitar in a few local bands, and worked with another in arranging the sound system. He was very busy with those enterprises, in addition to being in three music groups at college. He was indeed making a life for himself.

My husband and I did not see him very often, but cherished those times when we went to the college on occasional Sunday afternoons to watch him play in the jazz band and in the classical music concerts. There were some things, even if only a few, on which we were able to find common ground and on which we could relate.

Karen was advancing in her studies at the college and had selected English as her major. Ironically, in contrast to her experience at the high school level, her work was much esteemed. She received a special written commendation which was placed in her file, and an essay that she wrote, while there, was one of four, chosen by the faculty, to be published in that school's annual publication of scholarly work.

She selected courses in elementary and early childhood education as a minor in her program of study, thus including her enjoyment of children in her career endeavors. She tutored students at the college who needed extra help, and made good friends with the international student population. Her previous love of life, and for the people in it, was manifesting in a socially mature way. She continued to pursue her dance interests, managing to travel to her home city several times a week to take dance classes. She advanced to the point where she was choreographing and teaching some of the children's dance groups in preparation for the recital to be held at the end of the school year. She had matured.

During this time, she met a fine young man and began to date him steadily. He shared her zest, her depth, and belief systems. She was busy, but adhered to a health maintenance program, incorporating all that she had learned about herself and kept herself well! She was on a secure path at long last.

As for me, I was beginning to reap the rewards of the evolving scientific knowledge--about allergies, chronic fatigue syndrome, high triglyceride levels, and various nutritional theories. Research was beginning to be appear in articles and journals. I discovered and attended out-of-state professional seminars, gaining information which validated that my original hypotheses and treatments for my daughter's conditions were accurate. This, indeed was good news. Greater still was the fact that Karen's health and functioning were stable. She was finally diagnosed with asthma, the only other thing over which she needed to gain control, but she has done that, too.

In the meantime, I was getting positive responses to the resumes that I sent. I would have liked to have been employed in a position doing research relative to what I had learned, because I wondered how many other people had found themselves in a situation like mine. I wondered if there were other disorders which had been thought to be predominantly "psychological" in nature, that were accompanied by nutritional deficiencies, allergic conditions, or other biochemical problems. I wanted to use the knowledge that I gained and the compassion that I had always had to help heal those who, like my daughter and me, had been given labels which they did not earn.

I was interviewed for a few positions in which that might have happened. Financially, I was not able to wait to see if there would be a possible second interview for one research position and accepted the first job in which I had received confirmation of a hire. The position was as a case manager in a mental health system. My work experiences of nursing, counseling, working in, (and with), a variety of agencies and, particularly of having worked with patients/clients in their homes, was to my advantage. But it was thought that my lack of knowledge of the specific categories, symptoms and medications" for mental illnesses", did not qualify me to be hired under the title, "psychiatric nurse". I temporarily accepted that premise and remained in the non-nurse position for about a year, continually requesting that I be given a position with a "nursing" title and the salary to which I felt I was entitled. I emphasized time and again that, although I was not a "psychiatric nurse" as they knew that role to be, I brought a wide and unique dimension of nursing to my work. I told them that I brought a "holistic" approach. Some staff were recognizing that I brought some undetermined "special" aspect to my work (whatever that was), and acknowledged that. It was noted by an administrator, by my immediate supervisor, and by the psychiatric nursing supervisor who said to me, "you must be doing something right. Your kids are doing well, you're doing well, and your clients are doing well". They pooled their efforts on my behalf and soon after, I assumed a newly-created position as a nurse-clinician on their mobile treatment team. In this capacity, I worked with those clients who had been resistant to traditional methods of treatment, and also served as the first liaison nurse-coordinator for the homeless shelter--interviewing, assessing, and interacting with clients from the shelter, referring them to the center, as needed.

I observed clients closely as I worked earnestly and intensely with them. I noticed the courses, stabilities and relapses of the symptoms that fall under the labels referred to as "mental illness".

I saw numerous examples of, (what I considered to be), various biological, and environmental factors interacting with the psychological factors of those disorders and/or conditions known as "mental illness." I thought more about the B vitamins that I had read so much about, particularly the information relating to Niacin --(thought to be associated with a profound effect on brain function, with a deficiency of it having been known to produce the symptoms of dementia and psychotic-like conditions).[2] I did not do any formal studies of etiology. I was there to carry out nursing and case management activities.

I could not help but be impressed by the obvious potentiality of the would-be artists, musicians, poets and other kinds of talents that seemed to be inherent among that population. Many of the characteristics which literature has named as indicative of "gifted and talented" seemed to be widely present among these people. Many of them demonstrated a high and wide degree of sensitivities, social awareness, concern for others, strong-will, determination to control, and other characteristics. Unfortunately, though, for many of these people, those characteristics had not lent themselves to potentiality fulfillment. I could not help but wonder about the numerous "other" factors that must have been involved for them along the way, (other than the ones we label as "psychological" or "psychiatric"), which caused those clients to end up with those negative labels. We need to take a look "beyond the labels", I thought.

I worked long, hard hours in that agency. I made friends and further progress and truly "got on with my life." I had been living at home for quite awhile now and it was "O.K." At times I still felt a terrible void for what I thought we had "lost" in our family, but I carried on in a positive way knowing that crying over "should have beens" does not bring them back.

I guess that for all of us, there will always be the remembrance of some sort of "gap" in our lives during the difficult time of the break-up of our family. But I have always been a forward--looking person. I would recreate whatever positive aspects of that family life that I could. And occasionally, I could be found running from car to door and from door to car, hoping that my "run" muscles would begin to work.

The Blooms of Spring

Spring came. Karen was nearing the completion of her senior year in college, as well as preparing herself for the upcoming dance recital. One night after dance practice, she arrived home with several costumes. She picked out one, in particular, to show me. It was a beautiful soft white flowing gown, the one she would wear in the dance that she had choreographed for herself. As I admired the beauty of it, I happened to notice the label. It read, "Wind Beneath My Wings". In some strange way, our journey and our relationship had been affirmed once again!

It was a beautiful recital! Many of my co-workers and Karen's special friend, soon to be her fiancee, joined us at the performance and at a special gathering at my home. Love was returning to our lives once again.

I was satisfied that I had completed my goal of re-establishing myself in my profession and fully reactivated the credentials that would provide me with a sound basis for the roles which I knew I had to undertake. I purchased a computer and spent many hours getting acquainted with it. I used it to write letters to those who had been involved in the misdiagnoses and false perceptions of our family.

Our family is now free from the demands of "authoritarians" who tried so desperately to control our lives. We are free from the pressures of people "in authority" who define others and their value systems as negative if they are not identical to their own. Each of us is free to choose to live in a way that does not emphasize materialistic values, if we so desire. That, to me, is what life is all about,--following your own conscience, and your heart. We are free from a risk of having our children taken from us because we do not adhere to someone else's values and standards. But most important of all, we have gained the freedom to resume the difficult task of rebuilding our family unit without intervention. This challenge has been happening with positive results.

In the spring of 1991, just about 10 years after the onset of our "drama", a strange and wonderful event happened for me, which I share with you here.

The year after having read about the local Senior Olympics, I did decide to enter. I didn't pay much attention to the wording on the registration form that read for "master" athletes. Never having been an athlete, I really didn't know what "master" meant. I sent in my application anyway. About a week and a half before the event, I began to think seriously about it. I had never had gym classes, or even taken part in organized sports. I didn't even know the length of 100 meters, but that is what I signed up for--the women's 100 meter run. It was more like "me" to seek another small challenge and it was helping to move my spirit along in the healing process.

Sometime before the scheduled races, I brought my car to a nearby parking lot to approximate the distance I thought I needed to practice. I drove it for what seemed like a little less than 1/10 of a mile and made a mental mark of the distance. During the next few days, I went to that lot on my bike early in the morning. I quietly and privately practiced my "race". I listened to my body regarding footing, breathing, types of clothing and I learned some things about me. But, best of all, I learned that I thought I could do it.

Sunday, the day of the race, I awoke - eager to be at the 8 A.M. registration. I had been aware to eat the well-publicized "pasta" the night before and had a good breakfast before heading off for the University field. When I got there, people were arriving; some with families, some with picnic baskets and many wearing fancy athletic gear. I got my number and my T shirt and when I looked at that beautiful track, I wanted to try it. It was a nice one - bright, well-marked and nice and flat and I had never before run on a track. I asked someone nearby, "Can I run on that track?". When they said, "sure", I did a practice run and went home, happy and satisfied. There was one catch. I had not yet been in the competition. That was to be at 11:30.

When I returned to the field at that time, it was really alive! The Olympic flame was lit. People and activities abounded. Javelins were being thrown, names and events were being announced. I sat and waited until I was called for my group. I was to be in Line "C". I listened to the conversation around me. The other women spoke about other events in which they had run; other events that they were in today; whether or not we would use a "jack" to start. I began to realize the meaning of "master" athlete. They were so "seasoned' and here I was asking what line do we end on! I was told that they would tell me when to stop.

In my appointed place at the head of track #4, I waited for the starting gun to be fired. Just then, my ears caught the sound of the German oom-pa-pa band I knew I was ready! The gun fired and the women all took off at once ahead of me. They seemed to run in a horizontal line in front of me. But that didn't bother me. I ran my race. I ran proud and strong. When I neared what I thought to be the finish line, I slowed down only to be told "keep on running". (I really didn't know where the finish was)! Soon after, I was running through the flagged area. Proud of myself and my race, I stayed around to observe. A voice on the loudspeaker told us to watch for the postings at a designated location. When I looked at the recordings there, I saw that names had been placed in categories of three. My name was there under category #3!

While I continued to look at the chart, bewildered, wondering why my name was there, they called for those people whose names were on the list. As I went toward them, they motioned for me to stand on an Olympic-type staging (the high platform in the center and two lower stagings on either side). With the #1 winner already in the center, they beckoned to me to stand on one of the stagings beside her.

Amidst the revelry of music and applause, a medal was placed over my head. I had won the Bronze Medal! The women whom I had seen running ahead of me were in a different age category. I had won "The Bronze" in my age category. I was so proud!

It wasn't just that I had "run the good race"; they had made me a winner, besides. This was truly symbolic for me!

Home
(Emotionally-Physically-Spiritually)

In some strange way, I guess I do consider myself a "winner". Since the "adventure" with my daughter, I have formed many conclusions about life and our society. I have continued to seek and to read new information about the various conditions which my daughter experienced and for which we had no name. Through private discussions and conferences with researchers and lecturers in the field of nutrition and neurobiology, my hypotheses have been confirmed. There is an abundance of data written now about findings in the areas of anatomy, physiology, theology, nutrition, social psychology; about the theories of intelligences and environmental sciences.

I have experienced so much more fully how one fears what one does not know and tends to label it as negative, or to even proclaim that it does not exist.

I am thankful that I have returned to and confirmed my own sensibilities about the person whom I was meant to be, (with even greater responsibilities, it seems). I can no longer respond to negative labels or to the negativity implicit in people's voices in their attempts at condemnation. I have been that route. It also will not work for my daughter, for labels do not describe us. We moved through the labels and made well of our potentialities, I believe. I wish others the same good fortune. I have gained many insights during my journey. I share them here in the chapters which I have designated as *Lessons*.

Since my return home, I have been further healed by the many people with whom I have come in contact--people who were able to help me move to the next place in my life. Last year, I received notice that I was eligible to run in the National Senior Olympics, for which I am honored. Although I would have relished advancement and pleasure in that area of endeavor, my basic mission had not yet been accomplished-- that is to continue to reach, educate and help to heal (through the writing of this book), those who have experienced, (or may be prone to future experience) of the journey traveled by my daughter and I.

I am accomplishing those goals. I have reclaimed my professional status. I have reconnected with groups representative of my public health nursing experience, who have more than reaffirmed the positive professional identity which they bestowed on me several years ago. I have become active in the local alumnae organization of the university from which I received my Bachelor of Science Degree.

I have established a private nurse/counseling practice which serves as a base from which I can write, teach, counsel and advocate. I have given workshops to adults on the process of discovering and developing their unique, individual skills and talents which were not discovered and nurtured in traditional settings. I have accumulated nursing credits from courses which validate my original theories and, (independently-gained), knowledge about the relationship of nutrients to physiological, behavioral functioning and to some biochemical disorders.

I have realized the value and meaning of "non-traditional". I hope to bring that concept to new status with new meaning. "Non-traditional" is really a neutral term, and as such, has little or no meaning. The more positive term is "creatively gifted", a term which must become more honored and accepted, because we are all entitled to it--in the gifts of our own unique creation.

Most important to me, I have become a member of the newly-evolving American Holistic Nurses' Association. I am privileged to say that I have assumed the role of being its Coordinator here in my state. It my duty to help to educate the public on the concepts of holism, about energy systems, about our connectedness to one another and the need that exists to heal one another and our planet. I believe I can speak to that. I will watch it grow as I have watched my own children grow, with a sense of knowledge and accomplishment.

My son has recently completed his college requirements, following a longer course of study than most. He holds two Bachelor of Arts Degrees, one in Music and one in English. Through the last several years, he has combined theoretical learning with "hands-on" experience in the music world. A year ago, we visited him at Christmas time, where he was working part-time for the season. We watched small children gather around him, some waiting eagerly, some watching cautiously, while he gently beckoned them to sit on his knee. If the children who love him listened carefully, they could have heard in him (as we do), the distant sound of music, as can only be orchestrated and played through the skilled hands of that particular "Santa".

He continues to perform in bands, on both guitar and trombone and to work sound systems for some other bands, but his full-time employment at present is in restoring buildings and furniture. He is truly an example of one who uses the unique talents that were discovered and fostered in early childhood. As the small child who cooed at the trees, he remains close to nature. He is a fine chef. Seasonally, he eats fresh fish that he catches, cleans and cooks, and fresh fruits and vegetables from his backyard garden. He still reads encyclopedias. I wouldn't be one bit surprised to find that his I.Q. and achievement scores still range in the top two per cent in the nation. But the pursuit of that is not of our goal.

My daughter is a certified elementary teacher. She has been teaching for several months in an excellent school system, where they seem to fully appreciate her creative abilities, accomplishments and the values of character that she has maintained. She is pursuing a Master's Degree in Elementary education, and has one course remaining to complete that degree. (You can be sure that some of her courses were certainly relevant to the arts, creativity and "gifted".)

We have just rejoiced in the realization of one of many

accomplishments in her early career. She, along with five others, have just shared with their former English professor, in the co-authorship of an chapter appearing in a book published by the Modern Language Association. It serves as an innovative approach to teaching college students a course related to the study of female authors. But, most importantly, I think it speaks to the essence of the professor under whose tutorage, Karen's talents returned and thrived. In describing the nature of the article, and why she has presented it in dialogue form, the author-professor states "students' perspectives are as important as mine - more important, in fact, if teachers truly want to gain insight into teaching". She goes on to say, "I may still be the one who assumes greatest authority in both class and essay, yet I try to demystify my authority as well. Or, rather, not I but we - we demystify it."[5] What a magnificent role model for my daughter in her chosen profession!

Karen has maintained the grace and poise of her dancing skills, currently using them in a School for Healing and Martial Arts. There she learns knew skills and assists some of the small children in learning those same arts. She and her husband attend the school. It is wonderful to watch them engaged in the development of these disciplines and techniques and to see them "sparring" together. In her "spare" time, she and a friend make crafts for their small part-time business, (appropriately named), "Heavenly Handicrafts".

She married just eight months ago in a beautiful wedding ceremony. The service was held in that little church in which she and I placed the small white Poinsettia so long ago on Christmas Eve. So many of our friends from throughout the years were present and those who were not sent their love and support. Throughout the day, and long after it, people remarked about the "felt love and presence" that they had experienced. They referred to it by several names... "beauty", "class", "family", but all agreed that it was "special". One even referred to my daughter as a "vision" as she walked down the aisle. Perhaps Karen was a "vision" - a vision of the accomplishment of her song that she had composed so long ago-- "Changes for the Future."

The joyous sound of a brass quartet filled the Church. It added a special "touch" to the ceremony. I was very mindful of the music and the presence of the trombone player-- my son. My brother, the priest, and Mike's uncle, a deacon, complemented one another as they bestowed the blessings.

Karen was never anxious nor afraid. She was radiant, confident and most mature. Nor was she anxious when Mike flew them to a tiny island for their honeymoon. She relished in the adventure of the flight and of her part in "co-piloting".

I have done my job. She and Mike, together, will "touch the sky", literally and figuratively. She has her "wings" now. Mike has another kind - those of a licensed pilot--his hobby. I often think of him as Michael, the "archangel". He openly shows his care and respect for her. He knows what she has conquered and cherishes the beautiful woman that she has become. They are a wonderful couple, blending beautifully, enjoying both shared and individual pursuits.

My husband has retired now. He busies himself in the yard with his hobbies. He is fortunate in that he, his three brothers and sister, "Auntie Muffie", still enjoy good health. Most of them meet almost daily and carry on their tradition of helping one another in their tasks. They have kept a strong Italian family and tradition together. We look forward to many old-fashioned Christmases.

Not long ago, I had an opportunity to run with the official Olympic torch that was flown over the games in Athens, Greece. I carried it high and proud on the streets of a nearby city while television cameras captured this occasion. It appeared on the 6 o'clock news. What they did not know was that, for me, the carrying of that torch was symbolic. It was the light and the flame that I believe my daughter and I have lit and carried for many others in the hope that we have, at least, lit one small candle in our corner of the world.

Last, but not least, even as I conclude this story, a wonderful honor has been bestowed upon me. I have been invited, screened and accepted as a participant in the first-ever International Senior Olympic Games to be held in Bermuda in 1996. I am told that it is based on the World-Scholar Athlete Games, emphasizing creativity, sport, comradery, nutrition and fitness. I will run the Women's 100 meter dash and will compete in creative writing. I bring with me your trust, your confidence and your love.

Notes

1. Gennari, Elaine (Collins), "Little One", in *New Voices in American Poetry*- 1989, eds. of Vantage Press. New York: Vantage Press, 1989, p.132.

2. Mark, Vernon, H., Mark, Jeffrey, P. *Brain Power*. Boston. Houghton Mifflin Co., 1989. p.130.

3. Gennari, Elaine (Collins), "My Gift of Joy", in *New Voices in American Poetry*- 1989, eds. of Vantage Press. New York: Vantage Press, 1989, p. 133.

4. Carkhuff, Robert R. *Toward Actualizing Human Potential*. Massachusetts. Human Resource Development Press. 1981.

5. Clark, Beverly Lyon, Braun, Heather I., Dearing, Susan, Garvey, Mary-Beth, Gennari, Karen, Hemperly, and Henneberry, "Giving Voice To Feminist Criticism": A Conversation". In *Teaching Contemporary Theory To Undergraduates*, ed. Sadoff, Dianne,F. and Cain, William, E. New York:

   The Modern Language Association of America. 1994. pp. 125-126.

Panic Disorder

What do we know about this thing some call panic disorder, (with or without) agoraphobia and other terms in the category of symptoms. First of all, they are just that; "terms", names given to describe categories of symptoms. Since 1981, (when my daughter first exhibited some of the symptoms to which names "panic attacks" have now been given), scientists and researchers have explored what lies beyond these labels and have provided us with information and an understanding of what occurs beyond the labels.

Panic attacks are physical disorders. They are manifested by devastating episodes of panic that are sudden, unexpected and therefore, seemingly without cause. The reality is that there are physical causes.

The condition seems to run in families, more so in women and mostly occurring between the ages of 15-19. They differ from the simple phobias and the separation anxiety sometimes found in children of a younger age. The biological, physiological component can be triggered by physical illness itself or overwhelming stress, but the condition is finally recognized as a medical problem.[1,2]

The body's normal "alarm system", the set of mental and physical mechanisms that allows a person to respond to a threat - tends to set off unnecessarily, when, in fact, there is no danger. The resultant symptoms can take the forms of racing or pounding heartbeat, chest pains, flushes or chills, nausea, dizziness, perceptual distortions, terror - a sense that something unimaginably terrible is about to happen and that one is powerless to prevent it, fear of losing control and doing something embarrassing and even fear of dying. Breathing difficulty is a hallmark.[3]

The ones suffering from this conditions feel an urgent need to be free to escape and to find help. (Therefore the freedom to leave a place, if they are successful getting there is, of the utmost importance.) The attacks usually last for several minutes and are often recurrent. Because the attacks are sudden and unexpected, many become homebound, not knowing the cause and fearing a recurrence. Many who suffer from these attacks go on to develop agoraphobia, a long-lasting condition, characterized with repeated attacks and the constant fear of their recurrence.[4]

Treatments that have been found to be effective over a long range include medications, and desensitization, (a process whereby the person is gradually exposed to that which they fear), with a person that they trust. Although many schools of thought now believe that nutrition plays a very important role in these conditions, this still remains an issue of controversy for some.

Enid Peschel, who was co-director of the Program for humanities in Medicine; assistant professor (adjunct) of internal medicine at Yale University School of Medicine; one of the editors of Neurobiological Disorders in Children and Adolescents; and president of NAMI (National Alliance of the Mentally Ill), until her death in August 1994, gave lectures and presentations all over the country in her ongoing efforts to raise the awareness level of the nature of the neurobiological disorders, which include the anxieties and panic disorder.

I met Enid a few years ago at a meeting of the RI AMI (Alliance for the Mentally Ill). At that time she was trying desperately to introduce the term "neurobiological disorder as a replacement for the term "mental illness, which she says is old, non--scientific and prejudicial language.[5] We discussed my daughter's symptoms and condition and the "treatment" which I was made to undergo. Shortly afterward, I received a copy of the book, Neurobiological Disorders in Children and Adolescents, which she edited (with husband, Richard Peschel, professor of therapeutic radiology at the Yale University School of Medicine, Carol Howe, membership chair of the Children and Adolescents Network of the National Alliance for the Mentally Ill and James Howe, past president of NAMI). She urges us in the book and in the numerous presentations she has given, to accept the findings of the Neuroscientific Revolution, and to use scientifically-based language to describe disorders that are known to be physical, medical illnesses that are associated with malformations and/or malfunctions of the brain. These include anxiety and panic disorders. They are biologically-caused physical illnesses resulting in various cognitive, perceptual, emotional, and behavioral *symptoms*. It has been *documented* that the above are physical, medical illness: brain diseases and are neither "mental illnesses" nor "mental health" problems.[6] She further purports and emphasizes my plea--that professionals make accurate diagnostic distinctions between children with neurobiological disorders (NBD) as stated above and children with psychologically based problems, which are sociologically- or environmentally-caused problems - due, for example, to abuse, neglect, poverty, crime, and other such factors of nurture.

She tells us that "unfortunately, most of the institutions that interact with children and youth - including the health care industry, educational programs, legislative and governmental bodies, and the legal system - have not yet accepted the new scientific findings about the biological basis of many of these disorders, which are still categorized as "psychiatric" or "mental health "problems." "Thus most programs now in place largely ignore, or refuse to accept, the findings of the neuroscientific revolution of the last ten years."[7]

She states, that "until we can integrate the newest neurobiological findings into the fabric and consciousness of society as a whole, the revolution in neurobiology will remain only a collection of scientific facts, divorced from their logical implications in terms of medical, educational, and social actions. It is time for the institutions of our society to translate these revolutionary scientific findings into effective, reality-based actions."[8]

She worked tirelessly toward that end, establishing the Neurobiological Disorders Society, Inc., just prior to her death. Her husband, Richard E. Peschel, M.D., Ph.D. and Vice President of that society, has informed us, that she had "essentially completed the editing of the first four issues of "Neurobiological Disorders", the newsletter of the Neurobiological Disorders Society, Inc., and that they are devoted to her memory. He asks that we carry on her work, as she, also, had urged me to do, in "fighting for more equitable treatment" for patients with these problems. I am grateful to have the benefit of her research to add to the confirmation of the causes of my daughter's symptoms.[9]

I am also grateful to have a personalized note from her in regard to my writing of Beyond the Labels. It reads "Thanks for sharing your daughter's - & your - terrific success story with me." (December 1993.) She has sent me prints-out of terms and ideas which read "It would be good to get millions of people to write letters and make statements using material.......to make this message part of the national consciousness and agenda, and adds, "you--and everyone else--can use these words as your own."[10]

In that respect, I freely use, (but give appropriate credit to), the information which she has uncovered and disseminated for us through the newsletters of the Neurobiological Disorders Society, Inc. In the very first volume, she urges us to use the words of neurotransmitters (GABA, glutamate, norepinephrine, dopamine, serotonin) and other words such as membrane receptors and neuroanatomy.[11] She tells us that "approximately 1-3% of all children and adolescents suffer from a true anxiety disorder."[12] In this same volume of the newsletter, there is a review of an article appearing in *Neuropsychobiology,* "Changes in peripheral benzodiazepine receptors in patients with panic disorder and obsessive-compulsive disorder" by Rotondo A. Marazziti et al. We are told that "patients with panic disorder had a marked decrease in the number of DBZ receptors (compared to patients with OCD) and compared to normal controls." The article further states that "the BDZ receptors are very closely coupled to the GABA receptors (GABA=gamma-amino butyric acid) in the brain" and that "the GABA neurotransmitter system is known to act as an inhibitory system in the brain. Therefore, activation of the BDZ-GABA system via benzodiazepines will activate the GABA system which acts as an anti-anxiety system in preventing hyperexcitability and panic attacks".[13]

C. Norman Shealy, physician and author, has been studying neurochemicals and concepts of psychoneuroimmunology for many years. He reported on these topics at the fourth International Conference of the Psychology of Health, Immunity and Disease and the results of his studies also appear in a volume compiled by the sponsors of that conference, The National Institute for the Clinical Application of Behavioral Medicine, Shealy defines the neurotransmitter, serotonin as a monoamine which, among other things affects mood and sleep[14] and says that these levels of serotonin "tend to be low in patients with clinical depression."[15]

In 1989, a study done on 104 depressed patients found "one to seven neurochemical abnormalities in 92% of the patients." In the same individuals, 80% to 100% (depending on type of testing) were deficient in magnesium and every depressed patient was deficient in one or more of the essential amino acids.[16] He states that "without these building blocks, these patients cannot manufacture the neurotransmitters and neurohormones essential for mental well-being! "Thus nutrition and metabolism (digestion)/absorption) become critical factors in neurochemistry".[17]

In regard to fear, he reports that "the primary response to stress is fear...reactions to fear disrupt normal neurochemical homeostasis, leads to the fight or flight response, "which affects the biochemistry of the individual virtually instantly with major effects upon both neurotransmitters and neurohormones."[18] "If the stress response continues, during the state of adaptation both the neurochemicals (transmitter and hormonal) and various aspects of the immune system become increasingly maladaptive."[19] He states that "many of the neurochemicals measured are known to affect immune functions".[20]

He lists some of the therapies which hc reports are "are at least as beneficial as psychotherapy itself". Among them is nutrition. He says, "virtually, invariably these interventions will at least assist in the process of restoring psychological homeostasis. Psychotherapy will probably be less effective if neurochemical homeostasis is not assisted.

Psychotherapists should consider both neurochemical profiles and psychophysiologic homeostasic adjuncts in their practices."[21]

Ironically, while Karen was experiencing her mysterious symptoms, I had written a paper proposing a link between the inhibitory neurotransmitters, GABA and vitamin B6 as part of the reason for the compulsivity of addictions. I hypothesized that a link existed between a vitamin B6 dependency syndrome of genetic origin and the manufacture of chemicals and the carrying out of the chemical reactions necessary for the healthy functioning of a neurotransmitter system. I further linked GABA to be dependent upon B6 for its formation. Citing an article in Nutrition Reviews,[22] I quoted that "Pyridoxine dependency appears to be a familial disorder"[23] and that "the presumed genetic abnormality in pyridoxine dependency is an abnormality of an apoenzyme system in the brain, resulting in decreased formation of gamma aminobutyric acid (GABA) which is formed in the nerve cells from glutamic acid by the action of L-glutamic acid decarboxylase in the presence of pyridoxal phosphate, the active form of vitamin B6.[24] That was in 1983. Other research in the field of nutrition has established a theory of individual differences in nutrient requirements, based on evolutionary concepts.

Although I was not able to get it published in Nursing Journals, my daughter and I have benefitted from that research and from other facts that I was able to link together about us and neurotransmitter systems. We now visibly "see" the link between panic attacks, hyperactivity, GABA and vitamin B6. I have made note where our family history indicates a B6 deficiency in at least one instance. I have since made many links and have had many other verifications about the role of various nutrients in the making of the neurotransmitter substances.

Much is known now and still more still speculated about the workings of a person's neurotransmitter system and it is in this area, I believe, in which the therapist for the "gifted" sensed and stated that Karen had "hidden disabilities". Scientific research is now uncovering those mysteries. Researchers Judith and Richard Wurtman have done a vast amount of study now relating these chemical substances to dietary intake. In *Brain Power*,[25] Mark and Mark elaborate on these findings. They explain that three neuro-transmitters, dopamine, serotonin and norepinephrine affect the perception of sleep, mood, sensation, activity and aggression. "Their concentration in the brain can be affected by the intake of certain amino acids, (tryptophan and tyrosine), which are contained in food". They add that, "it is increasingly clear from research that brain messages passed from cell to cell by neurotransmitters can be enhanced by the food we eat". They report that the "effects that the Wurtmans and other brain researchers have observed about various dietary substances are real."[26] "Their work is at the cutting edge of new and important research on the relation between food, brain functions, feelings and behavior."[27]

There is still controversy surrounding nutritional therapy, but I have no doubts that nutritional therapy was, and remains to be, a significant factor in my daughter's life. Many practicing in the field of holistic nursing have long seen the results of some of the less widely accepted treatment modalities. But even mainstream nursing is finally beginning to recognize the significance of new findings and of the need for change.

Louise Nigh Trygstad, Professor of nursing at the University of San Francisco, School of Nursing, seems to recognize the current challenge to incorporate new methods and the findings of the neuroscientific revolution, into nursing practice. In an article appearing in *The Journal of Psychosocial Nursing*,[28] she says that "psychiatric nursing in the 1990's requires integrating an expanding biological focus into nursing practice to accommodate both the growing biological knowledge base and changing patients needs". She uses terminology reflective of that urged by Enid Peschel as she reports, "Congress labeled the 1990's the Decade of the Brain. The neuroscience knowledge expansion is reflected in the creation of such terms as neuroanatomy, neurophysiology, neurobiology, neuro-endocrine-immune system and neuro-pharmacology".[29] She writes of the need to update nursing programs to include "biological theories of panic, neuroanatomy of memory, circadian rhythm, and psychopharmacology and immunology."[30]

And Susan Reinhard, R.N.,Ph.D., realizing the need for families to increasingly participate in decision-making, writes of a view of caregiving which recognizes the value of families and professionals working together, recognizing and sharing knowledge. Writing in *IMAGE Journal of Nursing Scholarship*, she states that "as the professional relinquishes claim to exclusive authority, control is redistributed and families play an active role in decisions that affect them."[31]

In the introduction of Neurobiological Disorders in Children and Adults, Thomas Detre, president of the Medical and Health Care Division at the University of Pittsburg, writes, "The publication of this volume provides convincing evidence that psychiatry, accompanied by its two traveling companions, neuroscience and neuropharmacology - has returned from its self-exile into the mainstream of medicine and biology." "Gone too, are the days when responsibility for the patient's illness could be deposited on the family's doorstep with impunity. The "who done it" approach has been dislodged by careful studies on genetic vulnerability and investigations designed to reveal the biological mechanisms by which stresses and other environmental insults affect the developing organism in general, and the central nervous system in particular. Today physicians and other health professionals view family members as allies who can become effective partners in the treatment team only if they have been thoroughly educated about the patient's illness".[32]

I pray that this may come to pass and I hope the writing of this book has served this purpose.

Notes

1. Department of Health and Human Services, Public Health Service, National Institute of Mental Health. *Phobias and Panic*, 1986.

2. Department of Health and Human Services. National Institute of Mental Health, *Panic Disorder*, 1991.

3. Ibid.

4. Department of Health and Human Services, Public Health Service, National Institute of Mental Health. *Phobias and Panic*, 1986.

5. Peschel, Enid. ed. NAMI Can News, Co-director of the Program for Humanities in Medicine and assistant professor (adjunct) of internal medicine Yale University School of Medicine, Personal interview, December 1991.

6. Peschel, Enid, Peschel, Richard, Howe, Carol, W. and Howe, James, W., eds., *New Directions For Mental Health Services: Neurobiological Disorders in Children and Adolescents*, Joey-Bass Publishers, (San Francisco, Summer, 1992), pp. 3-5, 91, 92.

7. Ibid., p. 91.

8. Ibid., p. 92.

9. Peschel, Richard, M.D., Ph.D. Letter to author, August, 1994.

10. Peschel, Enid. Letter to author, December 1993.

11. Peschel, Enid, (ed), *Neurobiological Disorder Newsletter*, Vol. 1. No. 1. Fall 1994. p. 5.

12. Ibid., p. 4.

13. Ibid., p. 7.

14. Shealy. Norman, C., M.D., "The Neurochemical Substrate of Behavior", Fourth International Conference of the Psychology of Health, Immunity and Disease, 9-12 Dec. 1992, Hilton Head So. Carolina, in 4th International Conference *The Psychology of Health, Immunity and Disease*, Vol. B, The National Institute for the Clinical Application of Behavioral Medicine, Ct. 1992, pp. 437, 440..

15. Ibid., p. 440.

16. Ibid., p. 442.

17. Ibid., p. 442.

18. Ibid., p. 446.

19. Ibid., p. 446

20. Ibid., p. 452

21. Ibid., p. 447

22. "Pyridoxine Dependency", *Nutrition Reviews*, Vol. 25. No. 3. March 1967, pp. 72-75.

23. Ibid., p. 73.

24. Ibid., p. 74.

25. Mark, Vernon H., Mark, Jeffrey P., *Brain Power*, (Boston, Houghton Mifflin Company, 1989), p. 139.

26. Ibid., p. 139.

27. Ibid., p. 140.

28. Trygstad, Louise Nigh, RN, DNS., "The Need to Know: Biological Learning Needs Identified by Practicing Psychiatric Nurses", *Journal of Psychosocial Nursing*. Vol. 32 No. 2. February 1994, pp. 13-18.

29. Ibid., p.13.

30. Ibid., p. 18.

31. Reinhard, Susan C. "Perspectives on the Family's Caregiving Experience in Mental Illness", IMAGE, Journal of Nursing Scholarship, Vol. 26, Number 1 Spring 1994, pp.70-73.

32. Peschel, Enid, Peschel, Richard, Howe, Carol, W. and Howe, James, W., eds., *New Directions For Mental Health Services: Neurobiological Disorders in Children and Adolescents*, (San Francisco, Summer, 1992), p.5

Transition to Knowledge

Transition to Knowledge

More than 10 years have transpired since we began the journey related here. My daughter is well. She maintains the holistic health program that she devised for herself and has adhered to, (even since she married), and is functioning well. As a result of her experience, she has developed the sensitivities and knowledge that it takes to be a wonderful caring and creative teacher. My family is re-integrated.

I have gained an abundance of knowledge through the interconnections that I have made. I have accumulated more than enough research to validate and confirm my original hypotheses:

1. that Karen's symptoms were always those of the same biological, biochemical origin but were being manifested in different ways.
2. that those symptoms were physically-based, resulting in emotional, cognitive manifestations during an adolescent period of hormonal changes and increased stress
3. that they were dysregulations in her system and were associated with neurotransmitter functioning
4. that a malfunctioning neurotransmitter functioning had been present throughout her life and
5. that in some way dietary components could improve that functioning.
6. And that she did, indeed, have "hidden disabilities."

Hidden Disabilities

Thc name ("hidden disabilities"), has even appeared in recent professional literature. There is a whole group of professionals in the field of occupational therapy who work with those children.

One recent article appearing in their newsletter "The Sensory Connection", gives tips for the child who is "sensory defensive", described as "overly sensitive to and/or overly aroused by sensory stimulation including visual information, sound, touch and movement."[1] I was not at all aware of those definitions, nor did I find out until last year that there was a professional group dedicated to the care and training of these children.

I have kept pace with unfolding research in the field of neurobiology and with the newest findings of these as they relate to Karen's symptomatology. I have watched research unfold in the field of nutrition which provides information regarding how nutrients affect neurotransmitter functioning. Scientific literature is now documenting the connections and implications of the serotonin, norepinephrine, dopamine, GABA, as related to the function of neurotransmitter systems and nutrition research is providing us with information regarding how dietary content affects those same substances of neurotransmitters. The two fields of research, (as connected), seem not to have found their way into traditional medical practice yet. Perhaps I can help to make that link. My hypotheses have, for me, been proven without a "shadow of a doubt", as the saying goes.

Chronic Fatigue

A final reward comes to me from Dr. Jay Goldstein's research into chronic fatigue syndrome (CFS). In his book, Chronic Fatigue Syndromes: The Limbic Hypothesis, he suggests that the limbic system, (which is an older portion of the brain and the part associated with functions essential to the survival of the species), is "medicine's stepchild".[2]

Goldstein reports that "disorders related to it (the limbic system) have generally been labeled 'psychiatric', and not part of hard medical science";[3] adding that "using the contemporary methods of understanding health and disease, many physicians can make no sense whatever of the multiplicity of seemingly unrelated symptoms that comprise chronic fatigue syndrome (CFS). He writes about some of the symptoms connected to the dysregulation of this system, (the limbic system), and mentions the very ones experienced by my daughter at one time or another; chronic fatigue, disordered sleep patterns, inability to fall asleep even though exhausted, sinusitis with note of recurrent flu-like illness with sore throat and, even, panic disorder.[4] He further tells us that "cells that secrete dopamine, norepinephrine and serotonin are distributed throughout the brain stem and project to the limbic system"[5] and that "limbic dysfunction may produce the symptoms of chronic fatigue syndrome".[6] Once again, we seem to have an implication of at least the dopamine and serotonin systems that were so much a part of my personal research and the conviction of their relationship to my daughter's problems and dietary intake and utilization.

I have been able to find some comfort and comradery in his statement that "it is unfortunately an all too common practice that if a clinical situation does not fit into established modes of analysis, then it is not regarded as being worthy of serious consideration. Often such cases are referred to psychiatrists, being labeled, 'psychosomatic' or 'hysterical' with no further thought having been given to how such symptoms may have been generated."[7] "Another common way of dismissing a patient's complaints is by saying, 'I don't know what's wrong with you,-all your tests are normal'." Consideration may not be given to the possibility that the appropriate test may not have been ordered, or is not yet in existence."[8]

Analysis

It is no wonder that I was misunderstood and not believed. For the most part, that specific knowledge is just now making its way into the scientific communities. An article appearing in the Journal of the American Psychiatric Nurses Association, as recent as February 1995 describes the "complex of symptoms" of CFS;

"debilitating fatigue that worsens with physical activity, flu-like symptoms, immune dysfunction and cognitive problems and affirms that the name given to those symptoms, (CFS), was only officially introduced in 1988![9] (not until two years after my daughter graduated from senior high school.)

I am thankful for the confirmations that I have found and that the holistic treatment that I provided for my daughter was "right on", as they say. The research is out; the new ways are in, the blame and guilt are gone.

It has been said that "truth has two sets of values and both are needed to liberate the reality".[10] I believe that somehow, this episode was meant to be. Without it, I never would have had reason to liberate the truth of my reality. To do that, and to produce the accompanying lessons, we needed both the truth of that therapist and the truth of me.

I have read that an assumption has been made (by Kohlberg), that "there are two requisite factors for developing a moral understanding; "(1) the opportunity for an open, interpersonal interaction between people who are not more than one level apart and (2) a situation structured to provide some value conflict between those people."[11] I believe that those factors were present in the interaction between the therapist and I. For, during the course of therapy, we discussed the stages of moral development, (of which we both had knowledge), and there were many times that I was puzzled by that therapist's remarks. She made many comments that led me to believe that at times, she was withholding knowledge and at other times, seeing me as someone who acted according to an outside set of rules, a "perfectionism" that needed to be treated. The reality was, that I was operating according to my own inside set of rules, my spirituality

I had arrived at that goal of (communion), which religions teach is attainable, not by following some outside rules, but by experiencing the ebb and flow of life, absorbing and integrating its meaning for me. I had walked the difficult paths that brought me to that state of wholeness. And yet, the only way that I was able to arrive at the words of that journey, it seems, was to go away from them, to walk in the shoes of another's reality, and return with the words of my own truth.

I believe that as painful as the interaction with that therapist was, there was some reason that we met. Someone else's words penetrated my being. I was put into a box, defined and enclosed by her words, to be forever imprisoned by them if I could not break out of them. In that box, my sense of creativity, my humor and spontaneity, my song, my sense of self-acceptance, and of acceptance of others, (qualities given as descriptive of those who actualize their potential), (Carkuff)$_{12}$ were not only ill-defined, but defined as negative and characteristics to be "treated". I knew that I was not negative and that I had already found my truth.

But what of those who had not found their own truth and had those words directed at them? What if I had not already found those truths? Would I forever have had to live by someone else's truth? Do we allow the freedom to explore the truths of who we are? For, we are who we are...all different, with potential for physical and spiritual levels of development. Within us is that which is inherent to bring truths to their fruition. It takes courage and energy, but we do have the power to unlock the truths within us. I walked in the shoes of another to find that they were not mine, but, in doing so, they became mine, as well-- just like the young Indian who did not hear the final beat of music until he recognized his brother.

When I was away from home, in pain and in retreat, I wrote in my journal, "where is the music of my soul that must blend with the others?" I believe that I have found it again and with the knowledge of it, that fourth beat of brotherhood and universality.

During the writing of this story, I so often thought of a question that I remember from biblical passages (which I could not readily find, but remember from my early religious training)..... "and what did you do with the talents that I gave you"?..... I answer, "I hope I have used them wisely".

Notes

1. "The Sensory Connection", A Newsletter of Occupational Therapy Associates, P.C. Watertown, MA. Spring 1995.

2. Goldstein, Jay A. *Chronic Fatigue Syndromes: The Limbic Hypothesis*, (New York. London, The Haworth Medical Press, 1993), pp. 29,23.

3. Ibid., p. 23.

4. Ibid., pp. 39, 58.

5. Ibid., p. 23

6. Ibid., p. 23

7. Ibid., p. 40.

8. Ibid., p. 40

9. Anderson, Jill S. and Kayner, Dorothy. "Milieu Issues in the Treatment of a Person With Chronic Fatigue Syndrome on an Inpatient Psychiatric Unit". *Journal of The American Psychiatric Association.* Vol. 1. No. 1. (February 1995): p. 12.

10. Sonnier, Isadore L. (ed). *Hemisphericity as a Key to Understanding Individual Differences*. (Springfield, Illinois Charles D. Thomas Publisher, 1992), p. 18.

11. Carkhuff, Robert R., *Toward Actualizing Human Potential* (Massachusetts, Human Resource Development Press, 1981), p. 19.

12. Carkhuff, Robert R., *Toward Actualizing Human Potential* (Massachusetts, Human Resource Development Press, 1981)

Lessons

"Hidden Disabilities" or Individual Differences

(Their Meaning)

Even though the therapist for the gifted seemed to have a need to fit us into a view known to her, I must admit that, (to her credit), she recognized and admitted that there were other things that she did not know nor understand. She used a term "hidden disabilities", and, perhaps, unknowingly, very aptly. Because, to me, it is the individual differences that are inherent in all of us that make a difference--and also make a link--between us. But those differences are not always seen and often result in behaviors, which are not accurately categorized.

The important point that I wish to make here is that not everyone is the "norm" which others think them to be or wish them to be. We say that we know that and so "what is new?" But in everyday living and interactions, we really do not accept and act upon that fact of life (as I have just illustrated in this story). Society has "norms" into which they are constantly insisting that people should fit. But, not everyone fits into the same pattern. True, we are referred to as being in the category of man in the animal kingdom, but there are degrees of differences within that kingdom. Not everyone brings with them the exact same biological, biochemical intricacies, potentialities, or mechanisms. We have basic patterns for our evolution, which unfold in the process of our adjustment/adaptations. That takes place individually, at its own pace. But all of the institutions of Western society seems to thrive on a "norm", with everything else being labeled, often prefixed with dis-(diseased), dys-(dysfunctional, dyslexic), etc.

I think that it is important to remember that a "norm", (at least as described by Webster) is "the usual state, standard, model, or pattern", "corresponding to the median or average of a large group in type." Median is the midpoint of something, which automatically and necessarily implies others at both ends of the spectrum. These "others", (reaching towards both ends of the spectrum), are merely that...other ends of the spectrum--a variance of God's creatures to be understood and delighted in...just more of the various colors and patterns of which we and the rest of creation are made. Nowhere have I read that other ends of the spectrum were there to be changed, to be annihilated, to be squeezed into a thing called a "norm". Would you cut off the two ends of the rainbow spectrum or is it the total spectrum that lends us the beauty of the whole? So too, with humans. We need the totality of the spectrums of light.

There is a theory developing among some scientists, rapidly gaining in proponents, that the key to our behaviors can be found in our evolutionary history and that the characteristics we see in people were at some point in time necessary for survival of a particular species. It has been postulated that our differences and our "oddities" stem from our evolutionary ancestry.

We are 19th century people living in a body that houses some 3,000,000 genes that have traveled through our DNA system from the time of B.C. We are told that the latest level of the brain to develop is the cerebral cortex (responsible for our various human characteristics and adaptability), and that it appeared in its modern form about 50 million years ago. That is quite a long time ago and yet, little has really changed. Along the way, some of the abilities, which were at some time supposed to help us in the adaptation to our environment, have been dropped, some added, some changed, some lost. Not all of us have brought the same genes, preserved and exhibited in the same ways. The result is that we bring tendencies, dominances and other variances which present, change, and manifest in a wide variety of traits, propensities, and physical and behavioral characteristics along the way of our journey in life. You may have been endowed with some genetic predispositions that, through time, have become more readily adaptable to present day conditions than what I bring but that does not make you any more or less "normal".

It has often been said, what is this thing called "normal", anyway? To me that is the question with the resounding truth of there is no "normal"--just states of being along the way. But, not everyone, it seems, is able to view us in that manner. While there may be a beginning of acceptance of differences of outer obvious physical characteristics, (size, skin color, etc.), there is yet to be had that same acceptance of internal characteristics which are, for the most part, only now beginning to unfold and to be noted in scientific fields. It may be possible, and I believe probable, that one's outer physical characteristics are not the great differential or common denominator among us but that the true common denominator lies beneath the skin in the unfolding of our evolutionary genes. Science is now uncovering the basic biological, biochemical intellectual variances that have often remained hidden within us. The time may finally be approaching when those people, who for so long have been stating to "deaf ears", "but I am different" will not only be believed. but accepted in the full dimension of that difference.

In addition, to the characteristics comprising our physical component, we need to look at energy, (that with which we constantly interact), its basis, transformation and use. We know that sources of energy are available to us in a variety of ways, but we need to remember that we also vary in the way that we receive and are able to utilize those various sources of energy. Once again, we are not all alike even in this regard, although many medical "norms" would have us believe that. Research in the field of nutrition has established theories of individual differences in nutrient requirements, based on evolutionary concepts. Although they were proposed by some, many years ago, they are, only now, just beginning to gain in credibility and practical use.

Scientific findings are presenting us with data about so many of our "hidden" differences": differences in ways of utilizing food sources, which, in turn, is dependent upon the availability of various enzymes and/or the inherent ability to manufacture; differences in our inner anatomical structures, as noted by the use now of the PET and PECT scans; differences in the number and kinds of our various receptors and their distribution throughout our body; the way that we manufacture and use the chemicals needed for appropriate functioning of our neurotransmitter systems; differences in the way that we respond to light and the seemingly varying needs for it that exist among us; differences in the way that we respond to environmental stimuli, be it air quality, noises or other. Do we really take all of those factors into consideration in our everyday living or have we been taught to believe that those basic differences are instead, abnormalities, and should be put under a "dis", "dys" label? Can we remember the one common denominator that weaves its way among us---our spirituality, knowing that the rest is only our "casing"?

We need to acknowledge our individual differences more fully, I think in our everyday interactions and to recognize them as "normal" even though some may can cause inconveniences or even, difficulties. Nevertheless, they are inherited tendencies, not of our own choosing. We need to try to understand and accept that they exist, I believe, if we truly want to understand and accept our fellow man. Not to know about them, does not give us permission to be closed about listening to the knowledge of what constitutes another, and to think that because we are in the "norm", we are the only "O.K." and, therefore, have no need to try to understand others. This is not "O.K. For, each of us but a part of all of us.

I think of this reality so often as I view the ongoing struggles, between men and women; between what is called "straight" and "gay", forgetting that we possess varying degrees of the same source. To embrace one and neglect the other is to cause our own imbalance--and society's. Sometimes, I think that we fear that. We fear the unknown. We fear what it is that looks or acts "different", and so we label it as deficient, negative or other. We even proclaim that it is not "O.K." to have this or that characteristic, or maybe it is "O.K." that you're like that, but "not in my church" or my group. We need to dispel the myths. We need to look beyond the labels.

The time has come to illuminate society with a thousand lights. We can no longer destroy the potential gifts of those who, for years, have not fit into the molds that society wants or rewards. We can no longer place labels connotated as negative on those who "learn" differently, have same sex preferences, or even those who have fallen into the derelicts of our system because-- they are all part of us.

We have known that for years, but members of society still insist that we need to label people--for insurance purposes, for teaching purposes. Do we really? In many cases, labels are applied and left on people without bothering to recognize the innate differences as inherent and "O.K." We may temporary describe conditions of being, which are causing "dis" tress or "dis"-ease, (while we assist others to arrive at a more comfortable state of being), but, we must not leave people under labels. Instead, we need to concentrate on the behaviors that foster a greater state of being, acceptance and love. We need to look at people as they are...composites of elements, thoughts, innate biological, biochemical, structural differences. The differences are basic entities. What we call them, I believe, is what causes so many of society's ills. Can we treat differences in ways that honor and develop individuals? Can we describe behaviors in terms of actions, not in terms of names for people? Can we call people only by their birth names, knowing that any other name is a label conjured up to describe some behavior? And what about behaviors? What is the basis for them? Do we always know for sure? How many times have people defined others' behaviors, (without affirming with the other), if they are both placing the same reason on the purpose for the behavior, without even asking them what they are doing or why?-- I call that speaking for another, without their permission and, in essence, trying to control them by forcing them into your own knowledge base and perception. This has become rampant in our society!

I have asked these questions and presented these concepts as food for thought. We talk about bettering our society, but we, (so often), do not act on it. Some institutions even believe that more of the same, "the norm" is better. Surely not at the expense of the others. I think that we all need to work at changing our concepts of differences! We can do that by listening to people, by keeping abreast of all of the latest research as it presents itself to us and to remember to speak and to live a life that reflects that we really know and believe that we are all part of the whole, all part of the integrated One.

"Giftedness" and "Potential"

Left Brain/Right Brain
The Basis for a Holistic Society

What is this thing called giftedness? Is there really an answer for this? Some have thought that to be gifted was to be looked upon with awe, as with child prodigies, the "Quiz kids", "Whiz" kids and the great artists of the centuries. The term is usually applied because those people have demonstrated outcomes of their potential abilities. We actually see their gifts in the resultant verbal and artistic productions or reproductions. Individuals have utilized certain innate abilities or talents, bringing them into the fruition of gifts. Once again, I find myself looking at labels. Is the word really "gifted--or to have "potential"--and at some time, producing gifts.

I believe that the word "gifted" implies one who has already given <u>or</u> who holds the potential for a gift, then further develops that potential to the fruition of a gift/s for the good of society. But, most of us recognize that, in order for the gifts to be actualized, many other circumstances are involved. In addition to potentiality, factors such as nurturing, environment, motivation, unique sensitivities, and perhaps most important--innate biological, physical characteristics, are considerations.

Theories and concepts about this topic abound. Much has been written about the nature of giftedness and the associated concepts of intelligence and creativity. Attempts have been made to identify those who already excel in or to have the potential to excel in those characteristics. But, once, again, how does one really measure potential? It is really that which is yet, unseen. It, of necessity therefore, has to include all of those unseen physicalities of one's unique brain function and all that constitutes intelligence or the potential for it. Is this really possible? And yet, we continue to measure, label, and place people categorically according to something called "intelligence".

For a long time, standardized tests were used as the main criteria to determine intelligence, with much emphasis put on the results of the testing. One was told to select from the printed word, solutions or answers, predetermined to be the "best ones". This type of testing was considered by many to be an absolute reference point for the measurement of intelligence. Individuals even were named on how they scored on those tests. Those who scored below a determined standard were given a name indicating low intelligence. Those that scored above a certain number were given other names, such as gifted, or sometimes, even geniuses. They were capable of learning in a certain manner, were capable of expressing the desired knowledge and had the ability to "see" the same relationships of items or factors as were seen by testers or those designing the tests. Most fit the norm, so to speak and received their number. How one scored on an intelligence test was the "be all" and "end all" of who a person was, even though it did not allow for the various learning differences; it did not allow for variations in ability to express one's knowledge; and it did not allow an avenue to express a more expansive knowledge base. In other words, it did not take into account that others sometimes "saw" other answers and solutions to problems and for which there was no method or place in the testing procedure to indicate that. One was supposed to pick our only the "*best*" answers. "Despite their uses, I.Q. tests present a lopsided picture of any individual. They tell us nothing of creativity, initiative or emotional stability-qualities that can matter even more in life than sheer intelligence."[1]

For a long time, that type of testing was deemed to be "O.K." by and for many. It served some useful purposes. But through the years, new knowledge about learning, capabilities and various theories about the nature of intelligence became available, and along with it, a lot of controversy about the importance and significance of standardized intelligence tests. The expanded knowledge gave us indication that verbal and reasoning abilities are, by far, not the only kind of abilities that are important.

But, in spite of the new knowledge, and in spite of the limitations that we know are imposed through many standardized tests, many educators still define I.Q. as an ultimate standard for intelligence. Other educators and psychologists, however, have begun to act on the expanded knowledge base-- that there are other ways of knowing, of expressing and of being intelligent.

Some programs for the gifted seemed to take this additional knowledge into consideration. For many of those programs, the eligibility criteria included not only of having an I.Q. of 130 or more, but also included the potential for a high level of performance. Specific academic aptitude and general intellectual ability were still recognized as special, but so also now was creative or productive thinking, the visual/or performing arts, leadership ability and for awhile, psychomotor ability.[2]

Happily, the field of what was to be defined as "gifted", was expanding. The inclusion of the words "potential" and "creative thinking" in relationship to what is called gifted, inherently provided even greater opportunities for the development of a broader range of would-be gifts. To me this is a large undertaking and a wonderful contribution from the school systems which afford this opportunity. But, many, it seems, have provided these opportunities for pupils' potential development only in elementary grades. Unfortunately, gifted programs, for the most part, either do not continue or do not continue in the same form throughout the school years. And it is still often believed that students who seem to possess high potential will do well in a program that emphasizes acceleration. There are still many educators, seemingly at all levels of education, who still hold the narrow view that intelligent or gifted means high I.Q. and are even unaware of the character traits and needs of some others who are called "gifted". Therefore, I believe, we need to look "beyond the labels" at individual differences and potentialities.

Several people are now beginning to accept the theory of multiple intelligences developed by Howard Gardner of Harvard University's Project Zero. He says that "we celebrate children's many intelligences, not just in reading and mathematics, the traditional flags of exceptional ability, although these intelligences are certainly valued and nurtured. But there is also musical intelligence, bodily-kinesthetic intelligence (such as dance or fine motor skills), spatial intelligence (such as construction, assembly, or artistic skills), interpersonal intelligence (that combination of unique abilities to communicate well and get along with others), and especially intrapersonal intelligence (the child's recognition of his or her own abilities and interests)."[3] Each of these intelligences encompasses a wide variety of abilities within each category. I believe that this contribution by Gardner is no small gift for many in society. For now we can celebrate what we have always known-- that to help a friend, understand a rhythm, interpret an abstract, compose a song, bring an image to fruition, can be recognized as among the intelligences. Other knowledge is rapidly contributing to this theory and gaining in importance as we learn more about brain functioning; in particular the left and the right brains of the cerebellum.

For years it has been known that the brain has two cerebral hemispheres, each responsible for its own set of functions. The "left brain" has been thought to be, and even been termed, the logical, analytical calculating one, which tends to dominate personality.[4,5] "It specializes in language skills such as speech and writing, as well as in mathematics and reasoning".[6] It is additionally responsible for the tasks of right-hand control, written language, math and science skills and reasoning. For most people, the left hemisphere is dominant, for a few people, the right hemisphere is dominant.[7]

The right brain is "endowed with special powers of intuition and spatial perception" and "is particularly important to creativity, music, art and athletics."[8] It is in regions of this brain that the emotional aspects of language and communication, (governing and interpreting the emotional content of language, and coordinating the normal intonations, rhythms, and other emotional subtleties of human speech) is carried out. Other right brain (once thought to be the "minor" brain,) functions include: recognizing faces and shapes, recognizing tunes, appreciating and interpreting spatial relationships, left-hand control, art awareness, three-dimensional forms, insight and imagination.[9]

It is the right brain that grasps situations as a whole, unlike the logical, analytical left hemisphere.[10]

Nevertheless, there are two halves, right and left, and they "see" things differently."[11]

In addition to them being known as two halves, now we are told that everyone has two minds. The two brains, the left and right cerebral hemispheres, interconnected, together share the work of the brain, but each is capable of taking over the functions of the mind as a whole, even though they are not alike and have different roles in behavior.[12]

As a result of split-brain research studies, we are able to gain a new appreciation of right brain function. Research studies at the University of Chicago and at the California Institute of Technology, demonstrate "that the right brain, far from being a minor hemisphere, influences significant aspects of behavior."[13] and shows unexpected potential. Yet the left brain, until recently, was known as the major hemisphere and the focus of research, because of language being localized in that sphere. Unfortunately, "the nonspeaking right brain, although evidenced to have "special- and valuable-powers" was relegated to a back-seat role and called the mute or minor hemisphere.[14]

We have long known that some people think verbally and analytically while others think from a more intuitive focus, but somewhere, in the history of Western civilization, the left brain thinking became dominant and revered.

It has been written that "the emphasis on rational thought that began in the Renaissance has for centuries favored the left brain at the expense of the irrational right brain. And yet man's highest achievements seem to stem from the successful integration of both left-brain intellect and right-brain intuition."[15]

Research about the "split brain" phenomenon has given us some interesting insights. It is said that "nearly all discoveries in every field appear to involve a sudden right-brain inspiration. At an idle moment, maybe even in a dream, intuition makes an intellectual leap - sensing a solution to a long-standing problem, arriving at fresh understanding, reaching a new level of appreciation - in one quick burst of illumination. Then the left-brain intellect laboriously works out the details of this hunch, step, by step. Without the left brain's patient attention to detail, the right-brain hunch may remain fantasy."[16] "Albert Einstein, once remarked that his first inkling of a new idea often came to him by intuition - sometimes in visual images. 'The really valuable thing is intuition', Einstein said. "A thought comes and I may try to explain it in words afterwards."[17]

"Einstein's statement encapsulates the crucial role of each brain in the creative process. Without the right brain, there would be no idea; without the left brain, the idea could not be explained."[18]

D. H. Feldman, who has written about "developmental approaches to giftedness and creativity", perceives the concept of creativity as an "extension of knowledge". He says, "in its most powerful form creativity is the extension of a body of knowledge, beyond the structural borders that preceded its appearance. Thus, creativity in the sense of major contribution is built out of mastery of a field or domain."[19]

Others define creativity in a different manner. Albert and Runco, writing on "the achievement of eminence: model based on a longitudinal study of exceptionally gifted boys and their families"[20] state that "creativity is a blend of processes and values that motivate and infuse behavior, selective attitudes, and personality traits",[21] and "made up of a particular sensitivity and an orientation to problems, each of which controls our visions of self and opportunity. Together they permit - even require - that creative individuals have very personally determined, and at times perhaps even idiosyncratic, encounters with their environments and their self-images".[22]

They stress that "exceptional creativity, when it is present as a special capacity of an individual, shows up in active problem-finding and problem-solving efforts, as well as in the feelings of high responsibility and sensitivity to some pertinent aspect of one's career."[23] and that "only a few persons have the "right stuff" to show this creative potential consistently and significantly."[24]

But what is the "right stuff"? One theorist of human development, Dabrowski, writing on "Developmental Potential" incorporates a theory of what he calls "positive disintegration".[25] His concept of developmental potential is defined as "the original endowment which determines what level of development a person may reach under optimal conditions".[26] Included in his theory are five forms of overexcitability, "types of increased psychic excitability". The term overexcitability is referenced as "a special kind of excitability,...enhanced and distinguished by characteristic forms of expression", with each form being a "mode of being in the world"[27] -"modes of personal experience and personal action." They can be viewed as "channels of perception" through which information flows "in the form of sensations, feeling, experience, images, expectations, etc."[28] These channels, with special talents and abilities, are said to be the "components which constitute the original equipment with which a child enters life and which is later shaped by all the parental, peer, school and economic forces with which we are well familiar".[29]

If more than one of these channels have wide openings, allowing an abundance of images, sensations, etc., dissonance, tension and conflict can result. But these potentials can also lead to expansion and enhancement of mental development. Regardless of the amount of conflict and tension resulting, they are part of the developmental process of the development of potential from the beginning levels through to the highest level of "all-embracing love".[30]

It would thus follow that creative persons, (as Albert and Runco portray them), are "relatively open to their experiences, impulses, and imagery, mainly because they tend not to deny or overcontrol them."[31] It is this aspect of openness to experiences which affords an abundance of information upon which they can build and make connections. These authors also tell us that creative persons often are said to have the "capacity to express both masculine and feminine aspects to their personalities"[32] while reminding us that this "does not mean that creative men or women are more homosexual than less creative persons, but rather that they usually have a broader range of interests and, equally importantly, that they are motivated by a wider array of feelings than their less creative counterparts."[33]

Albert and Runco also address the widely-held notion about creative individuals being "non-conformists". They note that, while "in areas of life that are not central to their creative problems and interests, these people can be just as conventional as the next person,"[34] but "creative individuals' behavior is usually dictated by their own values (independence) rather than by others' (conformity)".[35] It is this very aspect of independence, that, I believe, prompts the labels of "rebels" and "non-conformists", that are often applied to them, even though the very nature of creativity implies non-conformity. Yet the reality is that they are "less concerned with social conformity and more interested than most persons in correcting or completing an otherwise unsatisfactory state of affairs."[36] They tend to be "philosophical", valuing problems and broad principles, with "a strong preference for dealing with theoretical issues and problems and an interest in and drive toward the aesthetic or elegant solution or experience".[37] These descriptions of creative persons closely parallel those attributes that Maslow applies to "self-actualizers".[38]

Robert Carkhuff, author of *Toward Actualizing Human Potential,* has written much about the nature of potential and self-actualizers. He describes creativity as a "universal characteristic" of self-actualizers and affirms that they are "ruled by the laws of their character, rather than by the rules of society".[39] He informs us that creative people are also spontaneous, accepting of self and others with all their shortcomings, but they also have "a strong sense of right and wrong, of good and evil.[40] They have a talent to discern accurately, and feel responsible to use their energy to solve problems in what they believe are missions. They are independent, tending to use their own resources and potentials. They have the ability to appreciate life again and again with naivetivity, pleasure and wonder, yet also have a need for solitude, sometimes experiencing the transcendence described as mystic experiences. Their relationships with other people are deep and meaningful with the capability for greater love and identification with others. He includes the idea of a spiritual potential, "a spiritual grace that accompanies human achievements"[41]. One of the missions of those who are "self-actualized is "to pass on the love of life - in all of its truth, beauty, and excellence."[42] Carkhuff summarizes them by saying that "in the end, actualizers value truth above all else"[43] and "realize that they must be their own pathfinders and travel roads never traveled before".[44]

One can only imagine what these people encounter in the course of their journeys. The characteristics described above are not always easy for the self-actualizer/highly creative person. The very qualities that lend to growth, movement and mission for that person, also leave him open to a barrage of criticism and loneliness. The end result is to choose to be true to oneself or fall into the status quo which is expected, and even demanded, from most of our institutions. It would not take much, particularly in adolescence, (when relationships are so important), to succumb to "the norm" and to "go the way of the crowd". It is even harder to follow one's path, if those adults doing the guiding have themselves fallen into society's "status quo" standards and so restrict that path. As a result, therefore, many never learn to understand themselves, feeling only that they are "different". The reality is that they are "different" and have to often live without gaining understanding, (much less), acceptance from others. These differences are often belittled; to have a vivid, active imagination is often frowned upon or ridiculed; to have spontaneity and naivete is considered to be childish and immature; intense helping relationships are often categorized as "dependent"; to have love without having to have sex is to be labeled as weird; to maintain the strength and courage of one's convictions is to be prone to be labeled as rebellious or non-conformist and to even be subject to disciplinary action for not conforming to a lesser moral standard. The traits of truth, child-like innocence and simplicity of spirit, traits which are often characteristic of the creative, are often condemned, in the name of "immaturity". Attempts are made by many to have them abort these characteristics in favor of an "adulthood" with lesser moral standards.

How difficult this must be for a potentially gifted child who is yearning and striving toward self-actualization? How do they fare or even survive in situations where others have a need to control? Some do not. How is this for children who have to function in a very structured school environment? It is the basis for crisis!

It is said that "it is in the social sanctions by peers against adolescent creativity that most gifted children lose their creative potential, and it is here that guidance for the gifted pays its greatest dividends."[45]

I have explained in detail the nature and functioning of the right brains and left brains. I have expressed the importance of the creative and intuitive aspects of right brain function. Yet, "most intelligence tests explore only the workings of the left half of the brain". It has been thought that left brain hemispheric function has inspired our Western, strongly verbal, scientific culture. This has happened while right brain functions, associated with the artistic, mystic cultures of the East,[46] and the hemisphere that grasps situations as a whole, has taken a "back seat"...to the neglect of the visual, performing and healing arts.

Society is in trouble. It needs the arts. It needs the healing arts. It needs right brain as well as left. If society is to benefit and promote self-actualization in people, it is necessary to allow for the development of both "schools " of thought.

Isadore L. Sonnier, Ed.D Professor of Science Education at the University of Southern Mississippi, is a stalwart in promoting the awareness and importance that both right and left hemispheres have in "collective functioning in society". He has edited a collection of works discussing this phenomenon. as it relates to all of our institutions of society. The essays reflect the contribution of the phenomenon of HP, (hemispheric preference), in fields of medicine, theology, government, social and family systems, management, engineering and, where all human interactions take place.[47]

He maintains that "hemispheric preference (HP) is the basis for the common thread of individual differences among human beings" and that these differences are "reflected in our collective differences and manifested, through social and political interactions, in all of our institutions.".[48] He calls on educational institutions, in particular, to be the initiators of healing and says that" in educational thought and practice, knowledge and understanding of the universal role of HP (hemispheric preference) has the potential for revealing not only how to reach and teach more students - but can also help us realize why we must do so."[49]

"Given the tenet that education is the main artery of sustenance for all other human institutions, holistic education is a test of its applicability. When both visual and analytical thought processing propensities of the hemispheric phenomenon are stimulated simultaneously and concurrently, it is tantamount to reaching and teaching both hemispheres, the whole mind, the whole person."[50]

In stressing the importance of utilizing the knowledge of hemisphericity in education, he writes of "the perpetuated misconception by an educational system which not only ignores the utility of visual functions and processes, but debates the validity of hemisphericity on the basis of flawed data".[51] He adds "perhaps we are short-changing ourselves when we educate only left-sided talents in basic schooling...the inverse relationship between scholastic achievement and creativity...the overtraining for verbal skills at the expense of nonverbal abilities."[52]

He urges us to consider and use this knowledge, not only in education, but in all of life's interactions suggesting that it is a "major contributor to individual differences in personality development among humans". He further asserts that "properly understood and implemented, this knowledge has the potential to remediate strife and disruptive confrontations in all of the institutions of human interaction."[53]

I could not agree more with Sonnier. I have written much in this book about the importance of considering all the intelligences, of respecting both right and left brain function. I think the real challenge, now that we are fully aware of the differences in brain structure and function, is to incorporate this knowledge in all aspects of society. If we really believe in improving human relationships and want to work at a more peaceful society, more understanding of one another, then we must use this knowledge toward that goal. We must remember the findings and the information given here about these potentials, yet unseen, yet the "stuff" of creativity. We must remember that creative means to build and not to be limited to what is already there. We must allow for expansion of our collective wisdom by listening to those who have it.

I believe that potential is manifested in behaviors at a very early age, whether they be in solving a problem, showing a knowledge of math or chemistry, singing a song, holding a hand or planting flowers. Those actions are showing the ability to function using the "stuff" we are made of, a variety of talent and more potential talent. Do not "snuff out" basic innate capabilities. Encourage the building of them--and the respect for them as the unique individual differences among us. No longer insist that someone be defined within a label, dependent upon an I.Q. result. Incorporate Sonnier's philosophy of a holistic method of education.

But, why do we really need to be told this? Haven't we always known that we have different biological, physiological inheritances from which we function and, therefore, relate? Why did I ever need to write about what we already know intuitively? Perhaps it is because those that do the judging, who have set the rules have been the more "verbal types", better able to express themselves in words, which move, enforce and control---words which have set the "tone" for Western society, often to the neglect of the other intelligences.

As I have pointed out, intuition, for the most part, has taken a "back-seat". If we really want to help form a more integrated, healthy society, I believe that the institutions within it need some revisions. We need the value of other tones-- the "tones" of the right brain, the tones of intuition and creativity - the colors, the sounds and the shapes, visually perceived and re-created. They too, have the power to move. They hold the energies of vibration in light, sound and thought forms. They are ultimately based on logic and mathematics, yet perceived by the senses. Through creation, they can be brought into existence and used productively. They can unlock words and "move mountains". They are the gifts of creation. We need all of them!

Notes

1. The Diagram Group, *The Brain A User's Manual* (New York. G.P. Putnam's Sons. 1982), pp. 113.

2. Shertzer, Bruce and Stone Shelley, C. *Fundamentals of Counseling*, (Boston, Houghton Mifflin, 1980), p.353.

3. Leibowitz, Deborah G., Starnes, Waveline T., "Unmasking Children's Gifts". *Gifted Children Today Magazine*. Vol 16, No. 5. (Waco, Texas, Prufrock Press. Sept/Oct. 1993), p.29.

4. The Diagram Group, *The Brain A User's Manual* (New York. G.P. Putnam's Sons. 1982), pp. 108-109.

5. Bailey, Ronald H., ed., "The Split Brain" *Human Behavior The Role of the Brain* (New York. Time Life Books. 1975.) pp.79.

6. Ibid., p. 79.

7. August, Paul Nordstrom, *The Encyclopedia of Psychoactive Drugs: Series II Brain Function* (New York, Chelsea House Publishers, 1988), p.90.

8. Bailey, Ronald H., ed., "The Split Brain" *Human Behavior The Role of the Brain* (New York. Time Life Books. 1975.) 1975, p. 79.

9. August, Paul Nordstrom, *The Encyclopedia of Psychoactive Drugs: Series II Brain Function* (New York, Chelsea House Publishers, 1988), pp. 86, 90-93.

10. The Diagram Group, *The Brain A User's Manual* (New York. G.P. Putnam's Sons. 1982), p. 109

11. Ibid. p. 109.

Notes (cont.)

12. Bailey, Ronald H., ed., "The Split Brain" *Human Behavior The Role of the Brain* (New York. Time Life Books. 1975), p. 79.

13. Ibid. p. 81.

14. Ibid. p. 80.

15. Ibid. p. 91.

16. Ibid. p. 91.

17. Ibid. p. 91.

18. Ibid. p. 91.

19. Feldman, D.H. ed. " A developmental framework for research with gifted children". In *Developmental Approaches to Giftedness and Creativity.* (San Francisco: Joey-Bass, 1982), p. 40.

20. Sternberg, Robert, J. Davidson, Janet, E. (eds)., *Conceptions of Giftedness* (Cambridge. New York. Cambridge University Press. 1986. rpt. 1990). pp. 332-357.

21. Ibid., p. 340.

22. Ibid., p. 341.

23. Ibid., p. 341.

24. Ibid., p. 341.

Notes (cont.)

25. Colangelo, Nicholas (ed.), *New Voices in Counseling the Gifted* (University of Iowa, Marquette University, 1979), pp. 25-55.

26. Ibid., p. 28.

27. Ibid., p. 28.

28. Ibid., p. 29.

29. Ibid., p. 29.

30. Ibid., p. 29.

31. Sternberg, Robert, J. Davidson, Janet, E. (eds)., *Conceptions of Giftedness* (Cambridge. New York. Cambridge University Press. 1986. rpt. 1990), p. 342.

32. Ibid., p. 342.

33. Ibid., p. 343.

34. Ibid., p. 343.

35. Ibid., p. 343.

36. Ibid., p. 344.

37. Ibid., p. 343.

38. Ibid., p. 342.

39. Carkhuff, Robert R., *Toward Actualizing Human Potential* (Massachusetts, Human Resource Development Press, 1981). p. 44.

40. Ibid., pp. 39-40, 43.

41. Ibid., pp. 39-46, 8.

42. Ibid., p. 130.

43. Ibid., p. 167.

44. Ibid., p. 166.

45. Colangelo, Nicholas (ed.), *New Voices in Counseling the Gifted* (University of Iowa, Marquette University, 1979), p.215.

46. The Diagram Group, *The Brain A User's Manual* (New York. G.P. Putnam's Sons. 1982), p. 109.

47. Sonnier, Isadore L. (ed)., *Hemisphericity as a Key to Understanding Individual Differences* (Springfield, Illinois. Charles D. Thomas Publisher, 1992) pp. ix, x, 6, 7, 9.

48. Ibid., p. 9, 6,

49. Ibid., p. ix.

50. Ibid., p. ix-x.

51. Ibid., p. 7.

52. Ibid., p.7.

53. Ibid., p. ix.

"The Battle of the Sexes" -
Does It Have To Be a Battle? One Woman's Perspective.

During the course of my parenting, I found myself amidst a myriad of societal views as to what should constitute male and female personhood. I listened and watched, but I had never had any problem with this issue. I was never a passive, subordinate female. In fact, my mother tells me that from the time that I could speak, I was very vocal; that, early-on, I informed her that some realities of life were wrong and that some things shouldn't be the way they were. I always made my goals, worked toward them and was usually successful in accomplishing them. The male hierarchy was never a problem for me except, perhaps, on the strict religious views that were attempted to be imposed upon me from my youth (which culturally stemmed from male thinking and imposition.) But I did not necessarily see those as male/female issues, but as "church" issues, which were to be logically debated and questioned. No, the woman's issue was never really one for me.

I followed some of my older brother's accomplishments, being the only one in my ninth grade class, (male or female), to enter independently the state science fair. And I won (for a project that I had created and designed). I did just about anything I felt I needed to do, at times though, with disapproval and reprimand. I was curious, innovative, aggressive and successful in my goals. I was a successful female, continuing through my early career into my marriage. For me, there was no male/female issue. But I watched many others and society struggling in an attempt to justify its perceived "wrongs" to women.

Shortly after my marriage, I observed the debates escalating and tempers flaring as increased demands were made by and for women in regard to their "rights". For several years there had been an increasing awareness and acceptance that woman is a person to be identified as that person in her own right. I agree that should happen. But it is an issue, I believe, to be looked at in perspective of history and science and reflects all aspects of society - social, political, economic and religious. It cannot just happen in a vacuum or overnight.

Traditionally, as we know in our country, men and women have had distinct roles in relationship to one another. For the most part, man was considered to be the achiever, the hard working, aggressive and assertive machine-like force and mover of society, while woman was thought (and often told) that she was to be his mate, but not implicitly his equal, in terms of economic worth, logic, etc. Traditionally a double sexual standard existed and was perpetuated. Sex was viewed as something to be sought after, and enjoyed by man, with woman to be the object of that sex. The Judeo-Christian tradition, characteristic of America, perpetuated these concepts.

The development and use of the birth control pill brought new sexual permissiveness in the 1960's and with it the onset of a drastic change in American social structure and a new dimension in sexual freedom. Women became free to plan child-rearing, leaving them more open to pursue careers. They saw themselves as liberal now. Where, for centuries, the strictly patriarchal hierarchy of the Church was accepted as such, it was, (and still is), being constantly challenged by women seeking their place, demanding to be ordained in major religious roles and sometimes being successful. The male/female issue even spread to the basic language of the bible, which began being labeled as "sexist". Big changes were occurring.

It took little for them to unite for their rights as they latched onto a movement led by Betty Friedan, the founder and first president of NOW, the National Organization for Women, "dedicated to the proposition that women first and foremost are human beings, who, like all other people in our society, must have the chance to develop their fullest human potential." Its stated purpose is "to bring women into full participation in the mainstream of American society now, exercising all the privileges and responsibilities thereof in truly equal partnership with men."[1] In her book, *It Changed My Life; The Women's Movement*, Betty Friedan writes, that movement "became known as a major movement for basic social change in this decade and possibly the most far-reaching revolution of all time". Further it was said to affect "our daily personal lives immediately, women, men, children; pervades all our institutions, office and home; confronts the economy, politics of right and left, theology, sexuality itself, in unpredictable ways."[2] And, indeed, it did.

From then on, society was never the same. Women demanded. Men became confused. Support groups sprung up for both groups to assist them in their changing roles. Women did begin to gain their worth and equality, which is important. That was all well and good, but all of those changes did not directly affect me at that time. I has always believed that women needed to be recognized and be given opportunities to grow, to achieve, to receive appropriate pay for the work that they do, to be allowed in positions of authority and leadership. But I had received that worth. That was no issue for me. I was an "at-home" Mom, willingly using my skills interactively with those of my husband. The key word, here, I believe, is willingly. That activity allowed me to restrengthen the skills that I had used in the work place and to use my intelligence in a productive way while parenting. In other words, I was using both masculine and feminine skills--logic, aggressiveness and initiative, while nurturing my children and providing a home and maternal love for all members of my family--a nice balance, I would say. I still held my old views about what I thought was a domineering church. But I watched as more and more woman became what, I thought, were combative, and even destructive, in their attempt to gain the entitlement that a gender role should bring. I watched, while yet another seeming imbalance developed as to how men should function and how women should get their rights. I did not always like what I saw. I questioned some methods and the rapidity with which they were striving to achieve these changes and I often thought, are they literally caught up in a "revolution", rather than working for an "evolution" of change?

In my opinion, men and women were still interacting as stereotypes, rather than according to each one's individual biological, anatomical, functional inheritance, (which really consists of both male and female traits and potentialities). To me there was little or no consideration of what it is that totally makes us male or female and how we each possess those inherent differences in perhaps different ways. The names, for the most part did not consider different bone and muscular structure with variations within each gender; they did not consider the left brain/right brain functioning as I have written--logical/analytical and artistic/intuitive with those varying within each sex; None of that seemed to be considered, much. Once again, all men are the same and all women are the same, the implication seemed to be. All men have the same "norms" for qualification and all women have the same "norms" for qualification. But no one really seemed to be looking at individual differences within the species, that there really are some "in-between" or "over-lapping" considerations, according to individual differences. Women just had to be women and to have, think, do and live the same way that a man does, regardless of those innate qualities within us that make us all different from one another and of individual differences *within* those labels of man and woman. Once again, even with the male/female issues, I thought, we need to look "beyond the labels". (For it was this very issue and one woman's strong feelings about what should be "right" for me as a woman that led to attempts at control in a situation which could have been a loving, relational, therapeutic one.)

I agree that all should be treated equally; that is that each person should be given equal worth for the person who they are and for the qualities that they bring to a situation, but that does not presume that we all function equally nor that we all want the same things out of life. Being allowed opportunities to which we normally should have been entitled does not mean that we all have to function in a specific way or that we have to pursue all of those opportunities if we are content with others. And to be content with a status quo that works for some, does not imply less femininity, less aggressiveness or anything else less. It means to be content within a certain situation, knowing that there are alternatives to that situation if we wish to choose them. We must recognize that there are varying degrees of functioning, abilities and desires in each human being, (regardless of whether they are men or women). What I believe should be the sole issue is that we are all equal in worthiness--to be appropriately affirmed; to receive benefits accordingly; to have opportunities, education, wages, housing, relative to the qualifications that we bring to a situation, and not merely because we are male or female.

But...I have been aware since very early in my lifetime, of the numerous rules that are in place to define and control individuals. Control is everywhere--in churches, in schools, the medical field, businesses, males, females, in so many of the so-called roles of "authority". Roles are used for control in so many areas. Rules, restrictions, intimidations and punishments seem to be the "key" to control. And it is not just a male/female issue. To me it, perhaps reflects one's own loss of control, countered by an attempt to regain it by trying to control others. If such were not the case, then I believe that we would simply have interactions with one another with shared knowledge. No-one would be trying to "control" anyone else with rules or labels.

I think that it is quite unfortunate for our human relationships that somewhere along the line in our history, women's opportunities were restricted in so many spheres. We have seen how the desire for power and control prompted dominance over another, setting up a battlefield. We should not let it happen again, by reversing the offenders and defenders. We should just continue to seek that which is rightfully ours. Power expresses itself, I believe.

But both sexes need to have sensitivity and awareness of our moral roles to the other in this controversy. Margaret Mead, noted anthropologist speaks well to this issue, I believe. She has said that "every adjustment that minimizes a difference, a vulnerability in one sex, a differentiated strength in the other, diminishes their possibility of complimenting each other, and corresponds symbolically to sealing off the constructive receptivity of the female and the vigorous outgoing constructive activity of the male, meeting them both in the end to a duller version of human life, in which each is denied the fullness of humanity that each might have had."[3] We need that "fullness of humanity" with all our interactions.

If we truly want to strengthen relationships and our society, we need to rethink this issue in light of knowledge about our inner male/female characteristics and potentialities, I believe. Eric Skjei, quoting Marc Feigen Fasteau, in The Male Ordeal,[4] gives insight into this possibility. "Perhaps in the future, our lives will be shaped by a view of personality which will not assign fixed ways of behaving to individuals on the basis of sex. Instead it would acknowledge that each person has the potential to be - depending on the circumstances - both assertive and yielding, independent and dependent, job and people oriented, strong and gentle, in short both "masculine" and "feminine"; that the most effective and happy individuals are likely to be those who have accepted and developed both these "sides of themselves." Can we once more, in another way, open the door to potential? Can we once more give credence to the belief that right brain and left brain traits exist in each individual and it is those that we must honor and respect, not just the word "male" and/or "female"? We can look "Beyond the Labels".

Notes

1. Friedan, B. *It Changed My Life; Writings on the Women's Movement*, (New York: Random House, 1976. xiv), p. 87.

2. Ibid., p. xiv.

3. Mead, M. *Male and Female; a Study of the Sexes in a Changing World.* (New York: Dell Publishing Co., 1969), p. 349.

4. Skjei, E. & Robin, R. *The Male Ordeal: role crisis in a changing world*, (New York: G.P. Putnam's Sons. 1981), p. 154.

Morality and Conscience
(What is This Thing Called Conscience?)

You have heard it said, "let your conscience be your guide". But is that really allowed? Do the dictates of Western Society really allow for that? Do present standards of society promote, (or even allow), advancement in the stages of moral development, (the "Six Stages of Moral Development"), a concept well-known in theological fields?

Lawrence Kohlberg, (probably the most well-known theorist in regard to those stages), believes that moral development is characterized by principles upon which a person bases his actions according to what he or she believes is right or just. They are said to occur in stages with each of six stages associated with a different social perspective.[1,2]

1. In the first stage, a person obeys some authority out of fear of punishment.
2. The second stage considers other people, sharing, and doing things for a reward.
3. The third stage goes beyond just the recognition of consideration for others. It includes concern for others in relationship, the majority is right and conformity is prized. Factors of trust, loyalty, respect and gratitude evolve. One supports rules which benefit the majority, and the majority is considered morally right.
4. The fourth stage emphasizes the group which keeps institutions going and for the maintenance of social order. There is concern that the rules be "right".
5. The fifth stage is based on the awareness that people hold a variety of values and is based on an internal commitment to principles of personal conscience, the concern that laws be based on the premise of "the greatest good for the greatest number."
6. The sixth stage centers on universal ethical principles, a kind of abstract morality which one believes to affect all beings regardless

of conventional views. When laws violate these principles of justice, human rights and respect for the dignity of human beings as persons, the person acts on these convictions with a sense of personal commitment.

It is said that "only about 10% of all people reach the last two stages of moral development"...and that "most gifted children do achieve these upper levels unless their environment has prevented such growth"[3]. In a book about Guiding the Gifted Child, we are told that "gifted children move more rapidly through those stages and that this "means that the gifted child is likely to be out of step with many of the needs, preferences, even personality traits of his age peers. He is, therefore, likely to question, challenge or defy traditions that his peers take for granted."[4]

Each of us has been given a whole set of mechanisms by which we are to survive, as "endowed" by our Creator. Left to our own resources, with the help and guidance of others and the forces of nature, we can develop to our fullest potential and hopefully achieve the ultimate holism inherent within us as we proceed through the six stages. But, it can be very difficult to do so when we are led astray, stopped on our path, or decide to stop on that journey...because we are alone on that path.

One reason that I write this story is to reach people. There are those who have various biochemical variances, environmental sensitivities, allergens, metabolic differences and a host of other "differences" not always understood by a society that has not experienced them. You can readily identify with this story. I ask you to take heart, to start anew on your journey.

To those who have realized your gifts, help others to do the same, but realize, while doing so, that you must acknowledge and allow for the variances that exist among us. For there are also those of you who have never experienced some of the variances, and who are in leadership positions. You have positions or roles of authority, and are looked to as those "authorities". You may work within systems that deal with issues of intelligence and/or learning, issues pertaining to physical matter, or moral/ethical issues. Many of you make rules, give orders, expecting others, including adults, to "obey" them. "Lack of obedience" is often accompanied by threats or punishments. Toward the accomplishment of what? Where is the allowance for the use of reason and for the acceptance of scientific variances, other than that which you know? Often, there is none. There are only hard and fast rules based on "norms" and known facts...with little or no exploration into the unknown. Often the very rules that are meant to help some, alienate others, from themselves and from one another. There are even some organizations--organizations of "religious" faith that have been founded to promote love--but in reality, restrict it. They proclaim to be universal in nature, but is universal love allowed within them? Can you belong if you are caring and intimate with a same sex partner--or are there rules against the expression of that love? --in essence saying, we care what your outer shell looks like--in preference to the spiritual being that expresses love--and regardless that research suggests that "sex roles may be as much a result of brain development as of social conditioning".[5]

I may have found the way to "oneness" and union, but can I be in com-union with your fellowship if I have been divorced and remarry? Is communion with one another strongly and consistently encouraged in your organization, (no matter what denomination or religious thought), in the knowledge that we are all one in the "Source"?

I know that organizations need rules and guidelines. But that is the very issue. Aren't they meant to be guidelines to the "Kingdom"? Why has such dominance, rules and punishment crept in? Would you forever keep me from reaching the kingdom that is within me? Is there a place for me if I follow my conscience, instead of your guidelines--or would you forever keep me at the first stages of moral development? In essence, "yes".

How far can onc get without being destroyed by the dictates of the "rules" or the "norms"? How far can one get if they have to "rebel" to break from the ties of society's "man-made" controls?" How far can one get when conforming to a conscience means to be called a non-conformist? How strong does one have to be in order to be able to find and follow their own "drummer"?

It is very difficult to march to the beat of the '"drummer" within us when it is different from most. Yet, I believe, in one sense we have no choice if some of us are to be true to our innate differences, if they happen to be other than the "norm". We may very well be different in our physiological bodily make-up. But, I believe that we all have an accompanying spiritual, energy body; that we have within us a natural "rhythm; a natural rhythm that corresponds and connects us with all of the other rhythms of the universe-- just as there is a "natural rhythm to the sun and the moon and the stars and the night and the tidal waves. We have to be in touch with that rhythm because we have to follow it. We have to "march to that drummer", even if it is different."

Therefore we who have other "norms" (and "different drummers"), need to disseminate information about us. We need to work for greater acceptance that biological, biochemical, anatomical structures, chemistry, physiology and other genetic characteristics are not all the same. Nor are all of our basic sensitivities the same (our receptors, our higher-sense perception). Although our worth is equal, the uniqueness of our creation is not the same. There are the differences with which our creator has endowed us. We are "fine" with our given characteristics, (those that we bring into this world, (our unique God-given endowments). It is man, I believe, that causes the problems, (spiritual ones), by telling us that what is innate, God-given, (in essence, our science, our conscience), is not fine-- and maybe, even needs to be changed. We need to know that we are the authorities of ourselves. We are our consciences (our beings based on our unique science).

I have heard it said that there is a seventh stage of morality. It encompasses all of the others and is called "why be moral?" Because we are a part of the whole and because of our universal connectedness to one another. The deeds of one person affect another person, in some universal manner. If we act immorally or fail to act morally, someone makes up the deficit. We draw from the universe and we give back to the universe.

There is an energy in us that we need to follow as it interacts with all that is in us and around us and in others. We need to be "in tune" with that God-given energy and make sure that we do everything in our power to keep it flowing in a positive direction. When possible, we need to diffuse and disintegrate that which is negative or destructive. To do so, we need to develop the awareness and sensitivities of our own selves and to those around us, to be in touch with, and to "hear" the "essence" within us, our God-given wisdom, the wisdom deep within us. It is this that I call conscience, meaning with science, guiding us, telling us what is "right". I believe it is the science of our matter and the connection of ourselves to all that lives, our larger conscience, "consciousness". It is that which is "right" within us, is distinct and unique for me, yet connects me to another. But how can I "hear" my inner voice when you override it. You tell me what is supposed to be or is "right" for me? How can I "hear" my own "guiding light", when you have darkened it and deafened it with a multitude of fallacies about who I am or who I should be or how I should learn or that only one intelligence is of value. How can I "hear" it when you tell me that you are my "God"?-- and even threaten me if I do not worship your word? The Bible tells me that the "Kingdom of God" is within me, but how can I reach it when you prevent me from doing so by making me abide by <u>your</u> "right", <u>your</u> "rules".

Like the Indian who beats the drum of which he knows, we must find the rhythm of our <u>own</u> souls, so that we may walk to it and with it and thereby, with one another. We need to return--to discover or rediscover that concept of "body wisdom" which should direct all of us to that which is "right" within us--the "spirit" within us--that which truly relates and connects us--sometimes referred to as the mystiquc, the mysticism that speaks to us, perhaps, more loudly than left-brained words and rules.

We need to let "spirit" evolve, I believe, without outside

interference or negative labeling of that which is inherently within us. We need to look "Beyond the Labels" if we are to find and truly relate to our own "higher" selves and therefore to one another. We need to let our spirits meet.

The ancient Indians seem to have discovered that secret as well as so many other facts of our existence, which they have expressed in myths or figurative language. Princess Redwing, "squaw sachem of the Narragansett tribes in R.I., (a descendant of Prince Philip, the Wampanoag leader and son of Woosamequin the dignified chieftain known to the Pilgrims as Massasoit - the Indian who kept the Plymouth Colony from starving in 1621) tells this story.

"I'm reminded"...."of a young Indian who stood on a hilltop with all the glories of nature around him. The young Indian heard a sound, a single drum beat - the sound of his own heart. He slept on the grass and drank from the brook, and the drum song became a two-beat cadence, to signify the union of man and Nature. When the young Indian sought an answer to the mystery of this union, a third beat was added: the sound of the Creator.

A fourth beat completed the drum song. The Indian did not hear this final beat until he recognized his brother.

"And that is why music the world over has four beats."[6]

Notes

1. Shertzer/Stone, *Fundamentals of Counseling*, (Boston, Hough ton Mifflin, 1980), p.53-55.

2. Webb, James T., Meckstroth, Elizabeth A., M.S. and Tolan, Stephanie S., *Guiding the Gifted Child*, (Ohio Publishing Co., Columbus, Ohio, 1982), p. 180.

3. Ibid., p. 179.

4. Ibid., p. 179.

5. The Diagram Group. *The Brain A User's Manual*. New York. G.P. Putnam's Sons. 1982.Putnam's Sons. 1982.), p.84.

6. Princess Redwing as quoted in news clipping (source, date un known) Notes

Conclusion

Conclusion

There will probably be those of you who will ask, "why did you write this story?" "Why did you so freely and openly express your feelings, thoughts and experiences and those of your family"? I, also have pondered those questions and have come to the conclusion that there was a need for this book and its contents, for it is more than just about raising two children. I believe there is a need, (for those who will look), to see the havoc that their need to control others can cause. There is a need for those who have been placed in roles of authority to know that they do not own the ultimate authority over another person or know all the answers. We know this, but the more that it is acknowledged and talked about, perhaps it will "catch on". There is also a need, I believe, for people to recognize the many facets of intelligence and spiritually. There is also a drastic need to lesson the attempts of those in authoritative positions to control, threaten and stereotype others. It should be "O.K." to question, to challenge, and to seek alternatives when existing institutions do not seem to be providing the appropriate solutions and/or guidelines. And those alternatives should be available, where possible. I do not believe in rebellion for rebellion's sake. I believe in reason for reason's sake. We need to address one another in reason, not in control, intimidations and threats.

I also think that there are many who would like to know that there still are people who will stand up for their beliefs and will challenge the "status quo", if necessary.

I hope that the reading of my story and the accompanying lessons are of some benefit, at least, to some people. The interactions which my daughter and I shared with others, serve to illustrate, I believe, the many facets of our humanity as we struggle to reach our ultimate wholeness.

Somewhere on my journey, I had found that wholeness. I describe it in the story as a feeling of oneness and said, "I knew that I belonged there...with all of nature." (p.55) But during the course of the so-called "therapy", I was given a choice, it seemed--to give up my daughter or to literally undergo a character assassination. I chose the latter.

I have never in my life, felt a need to "prove" myself. I was always, just "me" and I was "O.K." But this time there was more than just my daughter and I at stake. There were principles involved. There were others; those who had gone before us and would come after us and, who, perhaps, could not speak as well as I, or be able to reach others in the way that I might. I needed to disprove that faulty "diagnosis" and with that, prove other truths-- for them, as well as for me. I was none of those things of which I was accused.

I did not fit into those characterizations as stated. I was not the personality that was depicted for me. I had made a choice of motherhood, and all that the role held, with its inherent problems and joys. I was no wimp, dependent upon my husband as portrayed. I had a lot of love to give, which was continually replenished in me. I believe that love is called Agape. I found it abundant in nature. I saw it. I felt it. I received it. I had it to give to others. Not to do so, was to go away from my own personhood. But there was one more act of love for me to accomplish, it seemed. As a result of the trauma imposed on me, I temporarily went away from that feeling of oneness with myself and others. I felt as though I literally became dis-integrated, but then went on to achieve greater re-integration. I guess this needed to happen--for, I could not let, what to me was a high degree of spirituality, be condemned, due to others' lack of knowledge or understanding of it. What a trusting soul I had been. I needed to find the words.

It is said that the whole is greater than the sum of its parts. In my search for greater truth, I tried on the treatment of the therapist. To me, if I were ever going to find the words to describe, my view, my vision, obviously, different from hers, I knew that I would have to rule out her point of view and her words that had penetrated my being.

I needed to find new words. For words have power. They can define, inform, teach, encourage and more. But, they can also destroy, limit and stop--stop self-actualization. Words must not stop or block one's journey toward the ultimate potential.

And what is "potential"? To me, it is that which can be, but has not yet been realized or made possible. It exists in matter, energy, form, etc. It is simply "potential"; pre-existing; not yet made or made known; but available to be made into existing form or state, through visual, analytical or other means.

In the lessons accompanying *Beyond The Labels,* I have tried to portray something about the nature of this thing called, "wholeness" and potential and to give indication of the various notes that comprise it within us: the nature of individual differences; differences that are seen and unseen; differences between the known and unknown; the knowledge gaps, yet undiscovered, which I call "potential" because they contain information readily there, but not yet accessed into the understanding of reality. I believe that I have crossed one of the silent gaps wherein lies potential, and have converted that into knowledge.

I never would have been able to so fully describe the oneness of which I spoke unless I "gave it up"--unless I went through the separation and re-union, the disintegration and re-integration to find new union and with that, the resultant knowledge.

That all requires energy. It takes energy to reach that potential state of oneness with ourselves and with all of humanity and more energy to realize that we have known that state. I call that "potential energy", which is available in the universe to make "visions" become realized.

Vision is something that I think that Western society, and the institutions which govern it, do not have enough of. Its proponents claim that it values it. But I believe, that much of the existing vision that permeates our institutions is limited, therefore limiting people, so often forcing them to live within a limited view of reality, with its "status quo" orientation, and the mentality that one should not "rock the boat".

We live in a world where, all too often, the spoken word, proven science and predetermined logic is king. Much of what cannot be perceived and readily understood has, for the most part, been dismissed by Western society as nonsense, voo-doo, or some sort of mysticism, because it has not been explained to the satisfaction of some. In my opinion, our institutions have been too concerned with scientific "proof", ignoring, refuting, and making lesser of the "visual" or "felt" approaches to other aspects of truth, and giving little validity to experience.

And yet we know that there is another way of receiving, conveying and expressing that which is seen or "felt". It is in the tones, rhythms, colors and vibrations of nature.

We speak loosely of a sixth sense and we do have one; It is another "way" of "knowing". It is inherent within us and is, I believe, our intuition. There is something within us that is connected to every other living thing at the very core of our being, which connects us to one another in some form of energy system. I like to refer to it as our star of "being" or our star of light. I think that it is that universal intelligence within us and the potential "to know". It is to be found in every living thing and in every cell in our body. It is the stuff that binds us together. I believe we "know" it in our intuition.

Yet we have strayed far from the path of intuition, (that which can not be seen with the naked eye), although we know it exists and probably governs all things. We know that there is something "special" that guides all of nature, something that is inherent that tells flowers how to grow, bees how to pollinate, tides to ebb and flow and birds to migrate and to sing. I am sure that humans hold these same "instincts" in some form. Yet, many have lost the ability to follow intuition. We are constantly told who we are and who we should be, to the detriment of following our innate wisdom. And, yet it may be that intuition (that which has not yet been grasped), may, in reality be, the Absolute Truth, the highest potential, the "*n*th" degree, the highest mathematical equation, the greatest "logic" of all. We need to allow for the gifts of intuition.

The 1990's have been declared to be the decade of the brain. I believe that there will be new findings which will validate some of our heretofore, intuitive experiences and new knowledge which will validate that there are, (and will continue to be), other "ways" of "knowing" than through the five senses as we know them--and that it will include Higher Sense perception. Once this knowledge is grasped, acknowledged, and allowed its rightful place, I believe that mankind will have gained much.

For, current words (and the particular arrangement of them), can be limiting in describing what we "know". Perhaps true communication, (com-union), is found at the deeper level--in tones, in sounds, in "waves"-- in addition to the words. It is, perhaps, through these, that real "meaning" is conveyed.

Therefore, I urge all of us to pause with one another while speaking. Think--does there seems to be a misunderstanding? Am I on the same "wave length" about that which you speak? Are our words having the same meaning as the ones with whom we are relating? And do our words and actions match? We need to honor the gifts of both right and left brain. For unity is, after all, (it is said), based on the integration of polarities, of intuition and logic, art and science, male and female, words and actions.

Perhaps that therapist had some wisdom of which I had not yet experienced, when she spoke of the "therapy" being too painful. I do not know, for we did not "meet" in words. But I believe we "met" on some different wave length and that there was some reason for our meeting and that"reason" was to help me to find the words.

As I close this book, I returned to the journal that I had kept daily while I was taking Practicum II, the course in which I counseled my daughter. I vowed not to read it until I had completed this mission--this book. I looked for something that I knew I had written there at the time, but which I did not fully understand then. It came from a small clipping, (source unknown), about the Way of the Warrior Heart. It reads: "In the way of the Tineh (Apache), the warrior will do whatever is necessary for the good of the people and the whole of life. This becomes an open-giving. It is done regardless of personal response. The warrior heart will lean toward balance and harmony to bring about peaceful resolutions."

In the spirit of the warrior heart,..........Peace.

Further Readings for "Beyond The Labels"

Adams, Ruth and Murray. *Body, Mind and the B Vitamins*. New York: Larchmont Books, 1972.

Anderson, G.H. Hrboticky, M. "Approaches to Assessing the Dietary Component of the Diet-Behavior Connection". *Nutrition Review*. (suppl.): 42-50. May, 1986.

Anderson, Jill S. and Kayner, Dorothy. "Milieu Issues in the Treatment of a Person With Chronic Fatigue Syndrome on an Inpatient Psychiatric Unit". *Journal of The American Psychiatric Association*. Vol. 1. No. 1. February 1995. pp. 12-15.

Anderson, Linnea, Dibble, Marjorie V., Mitchell, Helen S., and Rynbergen, Henderika J. *Nutrition in Nursing*. Philadelphia: J.B. Lippincott Co., 1972.

August, Paul Nordstrom. *The Encyclopedia of Psychoactive Drugs: Series II Brain Function*. New York: Chelsea House Publishers. 1988.

Ayres, Jean, A. Ph.D. (with assistance from Jeff Robbins). *Sensory Integration and the Child*. 5th ed. California: Western Psychological Services. May 1982.

Bailey, Ronald H., (ed). *Human Behavior The Role of the Brain*. New York: Time Life Books. 1975.

"Biochemistry of the Water-soluble Vitamins," *Encyclopedia Britannica, Inc*. Vol. 19. 15th ed. Chicago: 1975.

Blair, D Thomas and Ramones, Valerie. "Psychopharmacologic Treatment of Anxiety." *Journal of Psychosocial Nursing*. Vol. 32. No. 7. July 1994.

Brownmiller, S. *Femininity*. New York: Linden Press/ Simon & Shuster. 1984.

Carkhuff, Robert R. *Toward Actualizing Human Potential*. Massachusetts. Human Resource Development Press. 1981.

Chopra, Deepak, M.D. *Quantum Healing Explaining The Frontiers of Mind/Body Medicine*. New York: Bantam Books. 1989.

Clark, Beverly Lyon, Braun, Heather I., Dearing, Susan, Garvey, Mary-Beth, Gennari, Karen, Hemperly, and Henneberry, "Giving Voice To Feminist Criticism": A Conversation". In *Teaching Contemporary Theory To Undergraduates*, ed. Sadoff, Dianne, F. and Cain, William, E. New York: The Modern Language Association of America. 1994. pp. 125-126.

Cohen-Holmes, Sandra and Land, Donald R. "Nutrition and the Dysfunctional Family". *Focus on Family*. March/April 1984.

Colangelo, Nicholas (ed.) *New Voices in Counseling the Gifted*. University of Iowa, Marquette University. 1979.

Connors, C. Keith. *Feeding The Brain*. New York: Plenum Press. 1989.

Considine, Douglas M., ed. *Van Nostrand's Scientific Encyclopedia*. Vol. 11. 6th ed. New York: Van Nostrand Reinhold Co. 1983. pp.431-432, 1949-2950

Department of Health and Human Services. National Institute of Mental Health. *Panic Disorder*. 1991.

Department of Health and Human Services. Public Health Service. National Institute of Mental Health. *Phobias and Panic*. 1986.

Drew, Elizabeth. "Four Faces of Able Adolescents", *Saturday Review*. 1963. 46 (3), p.69-71

Druyan, Mary Ellen. "Sleep, Diet and the Brain": Seminar for Health Professionals presented by Institute for Natural Resources. Framingham, MA. 21 July 1994.

Dhopeshwarkar, Govind A. *Nutrition and Brain Development*. New York: Plenum Press. 1983.

Edelson, Edward. *Nutrition and the Brain. The Encyclopedia of Psychoactive Drugs*. Series 2. New York: Chelsea House Publishers. 1988.

Farha, Bryan. "Nutrition and Mental Health." *Health Counselor*. Vol. 4. No. 6. 1992. pp.39-41.

Feldman, D.H. (Ed.), "A Developmental Framework for Research with Gifted Children". In *Developmental Approaches to Giftedness and Creativity*. Joey-Bass Publishers. San Francisco: 1982.

Fitzgerald, Frances E. "Healthy Digestion". *Health Counselor*. Vol. 4. No. 6. 1992. pp.23-27.

Francoeur, R. *Becoming a Sexual Person*. New York: John Wiley & Sons. 1982.

Fredericks, Carleton, and Herman Goodman. *Low Blood Sugar and You*. New York: Grosset and Dunlap. 1969.

Friedan, B. *It Changed My Life; Writings on the Women's Movement*. New York: Random House. 1976.

Gennari, Elaine (Collins), "Little One", in *New Voices in American Poetry*- 1989, eds. of Vantage Press. New York: Vantage Press. 1989, pp.132.

Gennari, Elaine (Collins), "My Gift of Joy", in *New Voices in American Poetry*- 1989, eds. of Vantage Press. New York: Vantage Press. 1989, pp.133.

Gershoff, Stanley. *The Tufts University Guide to Total Nutrition*. New York: Harper & Row. 1990.

Gibran, Kahlil. *The Prophet*. New York: Alfred A. Knopf. 1968.

Gold, Mark S. *The Good News About Panic, Anxiety and Phobias*: New York: Bantam Books. 1990.

Goldberg, H. *The New Male Female relationship*. New York: William Morrow and Company, Inc. 1983.

Goldstein, Jay A. *Chronic Fatigue Syndromes: The Limbic Hypothesis*. New York. London: The Haworth Medical Press. 1993.

Goodwin, Frederick."Molecular Genetics: New Window Into the Brain?" *ADAMHA News*. Vol IX. No. 10. June 1983. p.3.Grant, N.W. "Magnesium and Chronic Fatigue". Lancet 91. 338-66. July 6, 1991.

Guindon, A. "The Sexual Language; An Essay in Moral Theology" *Louvain Studies*. vol. vi. No.3. Spring, 1977. pp. 207-238.

Henley, Larry and Silbar, Jeff. "The Wind Beneath My Wings". Warner Brothers. 1982. 1989.

Janssens, L. "Norms and Priorities in a Love Ethics". *Louvain Studies*. vol.vi. No.3. Spring, 1977. p. 306.

Koomar, Jane, M.S. OTR/L. "Sensory Integration: Theory Update Making Sense of Sensory Integration." Newsletter *Occupational Therapy Association*. Winter 1993.

Krassner, M.B. "Diet and Brain Function." *Nutrition Reviews*. (suppl.): 12-15, May 1986.

Lama Anagarika Govinda. *Foundations of Tibetan Mysticism*. York Beach, Maine: Samuel Weiser, Inc. 1969.

Lieberman, H.R., Spring, B.J., et al. "The Behavioral Effects of Food Constituents. Strategies Used in Studies of Amino Acids, Protein, Carbohydrate and Caffeine." *Nutrition Reviews*. (suppl.): 61-19 May 1986.

Leibowitz, Deborah G., Starnes, Waveline T. "Unmasking Children's Gifts". *Gifted Children Today Magazine*. Vol 16, No. 5. Waco, Texas: Prufrock Press. Sept/Oct. 1993.

Linkswiler, Hellen. "Biochemical and Physiological Changes in Vitamin B6 Deficiency." *American Journal of Clinical Nutrition*. Vol. 20. NO. 6. June 1967, pp. 547-561.

Lovenberg, W. "Biochemical Regulation of Brain Functions". *Nutrition Reviews* (suppl.): 6-11, May 1986.

Marciniak, Barbara. *Bringers of the Dawn Teachings from the Pleiadians*. Santa Fe, New Mexico: Bear & Company Publishing. 1992.

Mark, Vernon H., Mark, Jeffrey P. *Brain Power*. Boston: Houghton Mifflin Company. 1989.

Markley, Elizabeth. Nutrition and Health: "The Brain-Body Connection: Hormones, Diet and Behavior": Seminar for Health Professionals presented by Institute for Natural Resources, Providence, R.I. April 29, 1993.

McGraw-Hill Encyclopedia of Science and Technology. New York: McGraw-Hill Book Co. Vol. 19 6th ed. pp. 256.

McGraw-Hill Encyclopedia of Science and Technology. New York: McGraw-Hill Book Co. Vol. 14 6th ed. pp. 490-491.

McMillan, C. *Women, Reason and Nature*. New Jersey: Princeton University Press. 1982.

Mead, M. *Male and Female; a Study of the Sexes in a Changing World*. New York: Dell Publishing Co. 1969.

Meister, Alton. "An Enzymatic Basis For a Blood-Brain Barrier? The Glutamyl Cycle - Background and Considerations Relating to Amino Acid Transport in the Brain." In F. Plum, ed, *Brain Dysfunction in Metabolic Disorders*. Research Publications: Association for Research in Nervous and Mental Disease. Vol. 53. New York: Raven Press. 1974.

Neurobiological Disorders Newsletter. Vol. 1. No. 1. Fall 1994.

Neurobiological Disorders Newsletter. Vol. 1. No. 2. 1995.

Neurobiological Disorders Newsletter. Vol. 1. NO. 3. 1995.

"Neurojuices May Regulate Pain, Memory, Need to 'Eat, Drink, and Sleep." In *ADAMHA News*. Callen, Kate, ed. VII. No. 21. Oct. 1981. pp. 5.

"News in Mental Health Nursing" "Doctors Use Neurofeedback Techniques to Treat Symptoms of ADD". *Journal of Psychosocial Nursing*. Vol. 31, No. 10. October 1993. p. 46.

On Being Gifted. By the American Association for Gifted Children. New York: Walker Publishing Co. 1978.

Ornstein, Robert and Sobel, David. *The Healing Brain*. New York: Simon & Shuster. 1987.

Peschel, Enid, Peschel, Richard, Howe, Carol, W. and Howe, James, W. eds. *New Directions For Mental Health Services Neurobiological Disorders in Children and Adolescents*. Joey-Bass Publishers. San Francisco: Summer. 1992.

Peschel, Enid. ed. NAMI Can News, Co-director of the Program for Humanities in Medicine and assistant professor (adjunct) of internal medicine Yale University School of Medicine. Personal interview. December 1991.

Peschel, Enid. (ed). *Neurobiological Disorder Newsletter.*

Peschel, Enid. Letter to author. December 1993.

Peschel, Richard, M.D., Ph.D. Letter to author. August, 1994.

"Pyridoxine Dependency." *Nutrition Reviews*. Vol. 25. No. 3. March 1967, p. 72-75.

Quinn, M.D. Matters of the Mind; Topics of Behavioral Neurology: Seminar for Health Professionals presented by Matters of the Mind Seminars, at Providence, R.I. May 1, 1993.

Reinhard, Susan C. Perspectives on the Family's Caregiving Experience in Mental Illness *IMAGE Journal of Nursing Scholarship* Vol 26, Number 1 Spring 1994 p.70-73.

Restak, Richard. *Receptors*. New York,Toronto: Bantam Books. 1994

Restak, Richard. *The Brain Has a Mind of Its Own. Insights from a Practicing Neurologist*. New York: Harmony Books. 1991.

Restak, Richard M. *The Brain: The Last Frontier*. New York: Doubleday & Co. 1979.

Russianoff, P. *Women in Crisis*. New York: London: Human Sciences Press. 1981.

Shertzer, Bruce and Stone Shelley, C. *Fundamentals of Counseling*. Boston: Houghton Mifflin, 1980.

Shealy. Norman, C., M.D. "The Neurochemical Substrate of Behavior." Fourth International Conference of the Psychology of Health, Immunity and Disease. 9-12 Dec. 1992. Hilton Head So. Carolina in 4th International Conference *The Psychology of Health, Immunity and Disease*. Vol. B, The National Institute for the Clinical Application of Behavioral Medicine. Ct. 1992.

Skjei, E. & Robin, R. *The Male Ordeal*: Role Crisis in a Changing World. New York: G.P. Putnam's Sons. 1981.

Sonnier, Isadore L. (ed). *Hemisphericity as a Key to Understanding Individual Differences*. Springfield, Illinois: Charles D. Thomas Publisher. 1992.

Spring, B.J., Lieberman, H.R., et al. "Effects of Carbohydrates on Mood and Behavior". *Nutrition Reviews*. (suppl.): 51-59. 1986.

Sternberg, Robert, J. Davidson, Janet, E. (eds). *Conceptions of Giftedness*. Cambridge. New York: Cambridge University Press. 1986. rpt. 1990.

Teyler, Timothy J. *A Primer of Psychobiology*: Brain and Behavior. San Francisco: W.H. Freeman and Co. 1975.

The Diagram Group. *The Brain A User's Manual*. New York: G.P. Putnam's Sons. 1982.

The New England Encyclopedia Britannica. Vol. 19. 15th ed. 1975. pp.488-489.

"The Sensory Connection", A Newsletter of Occupational Therapy Associates, .P.C. Watertown, MA. Spring 1995.

Trygstad, Louise Nigh, RN, DNS. "The Need to Know: Biological Learning Needs Identified by Practicing Psychiatric Nurses". *Journal of Psychosocial Nursing*. Vol. 32 No. 2. February 1994. (pp. 13-18).

Wallwork, James C., Milne, David B., Sims, Rodger L. and Sandstead, Harold H. "Severe Zinc Deficiency: Effects on the Distribution of Nine Elements (Potassium, Phosphorous, Sodium, Magnesium, Calcium, Iron, Zinc, Copper and Manganese) in Regions of the Rat Brain." *Journal of Nutrition*. Vol. 113 N0. 10. Oct. 1983. pp. 1895-1905.

Wasynczuk, Anne, Kirksey, Avanelle and Morre, Dorothy M. "Effect of Maternal Vitamin B-6 Deficiency on Specific Regions of Developing Rat Brain: The Extrapyramidal Motor System." *Journal of Nutrition*. Vol. 113. No. 4. April 1983. pp. 746-754.

Waterhouse, Debra M. Nutrition and Health: Mood, Mind and Appetite: Seminar for Health Professionals presented by Institute for Natural Resources: Hyannis, MA. July 17, 1992.

Webb, James T., Meckstroth, Elizabeth, A., Tolan, Stephanie S. *Guiding the Gifted Child*. Ohio: Ohio Publishing Co. 1982.

Weston, Lee. *Body Rhythm. The Circadian Rhythms Within You*. New York: Harcourt, Brace, Jovanovich Inc. 1979.

Williams, Sue Rodwell. *Nutrition and Diet Therapy*. 4th ed. St. Louis: The C.V. Mosby Co. 1981. p. 109-110.

Wurtman, J. Emotions, Overeating and the Brain: Seminar for Health Professionals presented by Walden Health Resources: Framingham, MA. February 1, 1993.

Wurtman, R. "Ways that Foods Can Affect the Brain." *Nutrition Reviews*. (suppl.): 2-5 May 1986.

Van Praag, H.M. "Neurotransmitters and CNS Disease". Lancet 1259-1264, Dec.4, 1982.

To order additional copies of **Beyond the Labels**, complete the information below.

Ship to: (please print)

Name____

Address______________________________________

City, State, Zip_________________________________

_____ copies of **Beyond the Labels** @ $12.95 each $______

Postage and handling @ $3.00 per book $______

Rhode Island residents add .91 tax per book $______

Total amount enclosed $______

Make checks payable to **Karma Publishing**

send to: Karma Publishing
P.O. Box 16103, Rumford, RI 02916

To order additional copies of **Beyond the Labels**, complete the information below.

Ship to: (please print)

Name____

Address______________________________________

City,State,Zip__________________________________

_____ copies of **Beyond the Labels** @ $12.95 each $______

Postage and handling @ $3.00 per book $______

Rhode Island residents add .91 tax per book $______

Total amount enclosed $______

Make checks payable to **Karma Publishing**

send to: Karma Publishing
P.O. Box 16103, Rumford, RI 02916